Lecture Notes in Mathemat

ISBN 978-3-540-50078-0 © Springer-Verlag 1

Angelo B. Mingarelli and S. Gotskalk Halvorsen

# Non-oscillation Domains of Differential Equations with two Parameters

## Erratum

p. 99, Postscript: Replace 1888 by 1988.

# Lecture Notes in Mathematics

Edited by A. Dold and B. Eckmann

1338

Angelo B. Mingarelli
S. Gotskalk Halvorsen

# Non-Oscillation Domains of Differential Equations with Two Parameters

Springer-Verlag
Berlin Heidelberg New York London Paris Tokyo

**Authors**

Angelo B. Mingarelli
Department of Mathematics, University of Ottawa
Ottawa, Ontario, Canada, K1N6N5

S. Gotskalk Halvorsen
Department of Mathematics, University of Trondheim NTH
Trondheim, Norway

Mathematics Subject Classification (1980): Primary: 34A30, 34C10, 45D05
Secondary: 34C15, 39A10, 39A12

ISBN 3-540-50078-2 Springer-Verlag Berlin Heidelberg New York
ISBN 0-387-50078-2 Springer-Verlag New York Berlin Heidelberg



Printing and binding: Druckhaus Beltz, Hemsbach/Bergstr.
2146/3140-543210

*... Et si illa oblita fuerit, ego tamen non obliviscar tui,*

*Ecce in manibus mei descripsi te.*

*( Is. 49, 15-16)*

**Per Felice, Angelino ,Dulcineo e Michelino**

*In memoriam*

## Preface

The aim of these notes is to study the large-scale structure of the non-oscillation and disconjugacy domains of second order linear differential equations with two parameters and various extensions of the latter.

We were heavily influenced in this endeavor by a paper of Markus and Moore [Mo.2]. The applications to Hill's equation, Mathieu's equation, along with their discrete analogs, motivated many of the questions, some resolved, and some unresolved, in this work.

As we wished to consider linear systems of second order differential equations, Sturmian methods had to be avoided. Thus we chose to base the theory essentially on variational methods - For this reason many of the results herein will have analogs in higher dimensions (e.g., for Schrödinger operators) although we have not delved into this matter here.

The first author (ABM) wishes to thank the Department of Mathematics, University of Trondheim - N.T.H., for making possible his visit there during September 1984 and for the kind hospitality rendered while there. He also wishes to acknowledge with thanks the support of the Natural Sciences and Engineering Research Council of Canada in the form of a research grant and a University Research Fellowship. Finally, he acknowledges with thanks the assistance of France Prud'Homme and Manon Gauvreau in the typing of various versions of these notes and his colleagues David Handelman and Edward L. Cohen for their generous time in technical assistance on the Macintosh SE.

Angelo B. Mingarelli

S. Gotskalk Halvorsen

Ottawa, March 1985 and May 1988.

# TABLE OF CONTENTS

## 1. Introduction

The study of the solutions of equations of the form

$$y'' + ( -\alpha + \beta B(x) )y = 0, \tag{1}$$

where $\alpha$, $\beta$ are real parameters and x varies over some real interval is of widespread importance in various branches of pure and applied mathematics. For example, the Mathieu equation, which arises naturally in connection with the problem of vibrations of an elliptic membrane, is of the form (1) with $B(x) = \cos 2x$. On the other hand, Lamé's equation, which occurs in the theory of the potential of an ellipsoid, is of the same type but with B(x) being a Jacobi elliptic function. Various eigenvalue problems may be cast in the form (1) - Fixing $\alpha$ and allowing $\beta$ to be the parameter we obtain a weighted Sturm-Liouville equation with a possibly sign indefinite B(x). Such equations have received much attention lately and we refer the interested reader to the survey [Mi.3] for further information on this subject. Of interest is also the case when $\beta$ is a function of $\alpha$ thus allowing for a nonlinear dependence on the parameter in question.

We are interested in studying the qualitative and spectral properties of equations of the form (1). In particular, we consider the set of all pairs $(\alpha, \beta)$ for which (1) has a solution which is positive in the interior of the interval under consideration and study topological properties of this set [defined below]. Central to our investigations is the paper by R.A. Moore [Mo1], who discusses the connection between the non-oscillation and periodicity of solutions of the Hill-type equation (1), in the case when B is continuous, periodic of period one and has mean value equal to zero.

Equation (1) is said to be **disconjugate on** $(-\infty, \infty)$ if and only if every one of its nontrivial solutions has at most one zero in $(-\infty, \infty)$. It is said to be **non-oscillatory on** $(-\infty, \infty)$ if and only if everyone of its nontrivial solutions has at most a finite number of zeros in $(-\infty, \infty)$. The collection of all $(\alpha, \beta) \in \mathbf{R}^2$ for which (1) is disconjugate (resp. non-oscillatory) on $(-\infty, \infty)$ will be dubbed the **disconjugacy domain** (resp. **non-oscillation domain**) of (1), and denoted by $D$ (resp. $N$). Moore [Mo1] showed that, in fact, $D = N$ and that $N$ is a closed, convex and unbounded subset of the $\alpha\beta$-plane which we call **parameter space** and label it $\mathbf{R}^2$ for simplicity. The results in [Mo1] were complemented by a paper of L. Markus and R.A. Moore [Mo2] in which B(x), appearing in (1), is now a (Bohr) almost-periodic function [Be.1] (or, for any sequence $\tau_n \in \mathbf{R}$, the sequence $B(x + \tau_n)$ has a subsequence which converges uniformly on $(-\infty, \infty)$).

The closedness of $D$ (and/or $N$) and its convexity are what we will term the large-scale properties of $D$ (or $N$) for reasons which will become clear below.

In this monograph we are concerned with the above-mentioned large-scale properties of $D$ (and $N$) for the more general equation

$$y'' + (-\alpha A(x) + \beta B(x))y = 0 \tag{2}$$

on the closed half-line $I \equiv [0, \infty)$ (although many of the results herein are also valid on $(-\infty, \infty)$). In (2) we assume **minimal requirements** on A, B, in general, in the sense that, A, B:$[0, \infty) \to \mathbf{R}$, and A, B $\in L_1^{loc}(0, \infty)$, i.e., they are Lebesgue integrable on every compact subset of $[0, \infty)$. Thus we waive all types of "periodicity" assumptions on A, B, and examine the consequences on the large-scale properties of $D$ and $N$.

It turns out that, in fact, $D$ is always a closed and convex set which can even be a **bounded** set in $\mathbf{R}^2$. In general $D \subsetneq N$ and $N$ is also convex though **not always** a closed set. (See §§ 2.1-2.2). In §2.3 we present general conditions on A, B which ensure that $D \subseteq N \subseteq H^+$ where $H+ = \{(\alpha, \beta): \alpha > 0\} \cup \{0,0\}$ as is the case for Hill's equation [Mo1].

The lack of periodicity-type assumptions on A, B usually has the effect of splitting $D$ and $N$, however we will see that $D = N$ may occur even in the "non-periodic" case (see § 2.1).

We then apply the foregoing results to the Sturm-Liouville equation (§ 2.5) and its extensions (§ 2.6) to potentials in which the eigen-parameter occurs quadratically.

In §2.7 we pose the general question - When is $D = N$ ? We show that, in particular, if A, B are Stepanov almost-periodic functions [Be1], then $D = N$ (§2.8). Further extensions of this result to the class of Weyl almost-periodic or more generally, Besicovitch almost-periodic functions appears doubtful, (see [Be1] for terminology).

In §3 we review the notions of disconjugacy for second order vector differential equations of the form (2) where, generally, A, B are n × n real matrix-functions whose entries are $L_1^{loc}(I)$. We introduce the new concepts of strong-and weak-disconjugacy and study the large-scale properties of $D$ in these cases. As is to be expected, when A, B are real symmetric, and "disconjugate" has its usual meaning, many of the results of §2 allow extensions to the vector case.

In §4 we analyze the large-scale properties of $D$ and $N$ corresponding to Volterra-Stieltjes integral equations [At1], [Mi2] as they include in their structure, the theory of differential equations and, furthermore, the theory of difference equations (in this case, three-term recurrence relations).

The techniques which allow these extensions are basically variational in nature, unlike the ones in [Mo1, Mo2] which relied upon variable change and the nature of B(x) in (1). In general one cannot rely upon Sturmian arguments. Thus it is possible, although we shall not delve into this matter here, to extend many of the results herein to the setting of elliptic partial differential equations with two parameters.

## 2. Scalar Linear Ordinary Differential Equations

In this chapter we introduce the sets $D$ and $N$, dubbed the disconjugacy and non-oscillation domains respectively. In the former case the equation (1) is disconjugate (i.e., no non-trivial solution has more than one zero in the interior) while, in the latter case, the equation is non-oscillatory (i.e., every solution has a finite number of zeros; note that the interval may be a semi-axis). Of particular importance is the study of the location of these sets in parameter space, and the associated geometry. The prototype, which serves as a basis for the cited study, is a Mathieu equation with a B(x) term which is periodic with mean value equal to zero. In this case, $D$ is contained in the right-half plane of parameter space- We seek to preserve this property by relaxing the assumptions on B(x). This leads to the following natural question: Which functions B have the property that the equation $y'' + \beta B(x)y = 0$ is oscillatory (on a semi-axis) for every real $\beta$ different from zero? We apply the resulting theory to the singular Sturm-Liouville equation $y'' + (\lambda r(x) - q(x))y = 0$ as well as to the equation

$y'' + (\lambda^2 p(x) + \lambda r(x) - q(x))y = 0$, which often appears in the physical literature. We note that weakening the periodicity-type assumptions on B generally has the effect of splitting $D$ and $N$. This, in turn, leads to the next natural question: When is $D = N$ ? Finally, we extend some of the results in [Mo2] to more general B's - namely those which are almost-periodic in the sense of Stepanov.

One of the key results in this chapter lies in the characterization of all those Bohr almost periodic functions B for which the equation $y'' + \beta B(x)y = 0$, has a solution which is positive on

$(-\infty,+\infty)$, for some real $\beta$ different from zero.

In this section we will always assume, unless otherwise specified, that A,B are real-valued, $\alpha$, $\beta$ are real parameters and that A, B are locally Lebesgue integrable over I (generally taken to be either $[0,\infty)$ or $(-\infty,\infty)$). The notions of disconjugacy and non-oscillation in the case of the half-line I are similar to the one given in §1, (see e.g. [Ha3]). Hence we can define $D$ and $N$ in this case analogously for (2).

### 2.1 The disconjugacy Domain

In the sequel it may be helpful to view the parameter space $\mathbf{R}^2 = \{(\alpha,\beta): \alpha, \beta \in \mathbf{R}\}$ as a linear vector space over $\mathbf{R}$ with the usual operations of vector addition and scalar multiplication. The functions A, B are **linearly dependent** if there exists constants a, b $\in \mathbf{R}$, $a^2 + b^2 > 0$ such that $aA(x) + bB(x) = 0$ a.e. on I. A subspace S of $\mathbf{R}^2$ is said to be **proper** if $S \neq \mathbf{R}^2$ and $S \neq \{(0,0)\}$. It is clear that all the proper subspaces of $\mathbf{R}^2$ consist of full rays through the origin (0,0) in $\mathbf{R}^2$.

The symbol **AC[a, b]** (resp. **$AC_{loc}$(I)**) will be used to denote the **class of all real-valued functions which are absolutely continuous (resp. absolutely continuous on every compact subinterval of I) on [a, b].**

For a given compact subinterval $[a, b] \subset I$ we define a vector space, with the usual operations.

$$\mathcal{A}_1(a,b) = \{\eta \in AC[a, b]: \eta' \in L_2(a, b), \eta(a) = \eta(b) = 0\}$$

Now let $q: [a, b] \to \mathbf{R}$, $q \in L_1(a,b)$. Then for $\eta \in \mathcal{A}_1(a,b)$ the functional

$$I(\eta, q; a, b) = \int_a^b \{(\eta'(t))^2 - q(t)(\eta(t))^2\}dt \tag{2.1.1}$$

is defined.

The first lemma is classical.

**Lemma 2.1.1** **Let $q: I \to R$, $q \in L_1^{loc}(I)$.. Then the equation**

$$y'' + q(x)y = 0, \quad x \in I \tag{2.1.2}$$

**is disconjugate on I if and only if for every closed bounded subinterval $[a, b]$ of I, the functional $I(\eta, q; a, b)$ is positive-definite on $\mathcal{A}_1(a,b)$ i.e., $I(\eta, q; a, b) > 0$ for each $\eta \in \mathcal{A}_1(a, b)$ and equality occurs if and only if $\eta = 0$.**

Proof. A proof in the case when $q \in C(I)$ may be found in [Ha.4]. The general case follows an analogous line of thought. For a more general result see §4.

We will use lemma 2.1.1 in order to prove lemma 2.1.2 which is central in this subsection.

**Lemma 2.1.2** **Let $q: I \to R$, $q \in L_1^{loc}(I)$. Then the equation**

$$y'' + \lambda q(x)y = 0, \quad x \in I, \tag{2.1.3}$$

**is disconjugate on I for each real value of $\lambda$, $-\infty < \lambda < +\infty$, if and only if $q(x) = 0$ a.e. on I.**

Proof. The sufficiency is trivial since $y'' = 0$ is certainly disconjugate on I.

In order to prove the necessity of the condition, let (2.1.3) be disconjugate on I for every $\lambda \in \mathbf{R}$. Lemma 2.1.1 now implies that whenever $\eta \neq 0$, is in $\mathcal{A}_1(a,b)$ where $[a, b] \subset I$,

$$I(\eta; \lambda q; a, b) > 0 \tag{2.1.4}$$

for every $\lambda \in \mathbf{R}$. This said, let $[a, b]$ be a given subinterval of I and fix $\eta \not\equiv 0$ in $\mathcal{A}_1(a, b)$. Since $\eta' \in L_2(a, b)$ it follows from (2.1.4) that

$$\lambda \int_a^b (\eta(t)^2) q(t)dt < \|\eta'\|_2^2 \tag{2.1.5}$$

where $\|\eta'\|_2$ is the usual norm on $L_2(a, b)$. We emphasize that (2.1.5) is valid for **every** $\lambda \in \mathbf{R}$. Since $|\lambda|$ can be chosen arbitrarily large it follows from (2.1.5) that

$$\int_a^b (\eta(t))^2 \, q(t)dt = 0 \tag{2.1.6}$$

Hence (2.1.6) holds for every $\eta \in A_1(a, b)$. Since $[a, b]$ is also arbitrary we find that (2.1.6) holds for every $\eta \in A_1(a, b)$ and for every $[a, b] \subset I$.

Now the test function $\phi_\varepsilon(t)$ defined by

$$\phi_\varepsilon(t) = \begin{cases} t-a & a \le t \le a+\varepsilon \\ \varepsilon & a+\varepsilon \le t \le b-\varepsilon \\ b-t & b-\varepsilon \le t \le b \end{cases}$$

is in $A_1(a, b)$, for each $\varepsilon > 0$. Inserting this in (2.1.6) and passing to the limit as $\varepsilon \to 0^+$ it is easily seen that (since $q \in L_1(a, b)$),

$$\int_a^b q(t)dt = 0 \tag{2.1.7}$$

Thus (2.1.7) holds for every compact subinterval $[a, b] \subset I$. Hence $q(t) = 0$ a.e. on I and this completes the proof.

**Remark 2.1** Since lemma 2.1.1 is valid for disconjugacy on intervals other than a half-line, it follows that in the hypotheses of lemma 2.1.2 one may replace I by $(-\infty, \infty)$, $(a, b)$ etc. (see e.g. [Ha.4]).

An interesting formulation of lemma 2.1.2 is its contrapositive.

**Corollary 2.1.3 Let $q: I \to \mathbb{R}$, $q \in L_1^{loc}(I)$ and let $q(x) \neq 0$ on some set of positive Lebesgue measure contained in I. Then there exists at least one value of $\lambda \in \mathbf{R}$ such that (2.1.3) is not disconjugate on I.**

**Corollary 2.1.4 Let $r_i: I \to \mathbf{R}$, $r_i \in L_1^{loc}(I)$ for $i = 1, 2, ..., n$. If the single equation in n-parameters**

$$y'' + (\lambda_1 r_1(x) + \lambda_2 r_2(x) + \cdots + \lambda_n r_n(x))y = 0 \tag{2.1.8}$$

**is disconjugate on I for every point $(\lambda_1, \lambda_2, ..., \lambda_n)$ in $\mathbf{R}^n$, then $r_i(x) = 0$ a.e. on I for $i = 1, 2, ..., n$.**

**Remark 2.2** Note that, once again, the contrapositive of corollary 2.1.4 is of interest - Thus if none of the functions $r_i(x)$ vanishes identically (i.e., a.e.) then there exists at least one point $(\lambda_1,...,\lambda_n)$ in $\mathbf{R}^n$ for which (2.1.8) is not disconjugate on I.

**Theorem 2.1.5 Let A, B as in the introduction to this section. Then the disconjugacy domain *D* of equation (2) is the whole space $\mathbf{R}^2$ if and only if A(x) = B(x) = 0 a.e. on I.**

Proof. This is basically corollary 2.1.4 for an equation with two parameters, the proof of which is straightforward (since we may set $\lambda_j = 0$ for all j except one and apply lemma 2.1.2).

**Corollary 2.1.6 If at least one of the functions A(x), B(x) is different from zero on a set of positive Lebesgue measure, then *D* is a proper subset of $\mathbf{R}^2$.**

Proof. This is the contrapositive of theorem 2.1.5.

The question of the boundedness of non-boundedness of *D* is a difficult one. - The next result gives a necessary and sufficient condition for *D* to contain a full ray through the origin of $\mathbf{R}^2$, and thus a sufficient condition for the non-boundedness of *D*.

**Theorem 2.1.7 *D* contains a proper subspace of the vector space $\mathbf{R}^2$ (other than the subspaces $\alpha = 0$, $\beta = 0$) if and only if the functions A, B are linearly dependent over I.**

Proof. Let A, B be linearly dependent. Then there exists $c \neq 0$ in $\mathbf{R}$ such that A(x) = cB(x) a.e. on I. Equation (2) then becomes

$$y'' + (-\alpha c + \beta)B(x)y = 0.$$

Hence *D* contains the subspace $S = \{(\alpha,\beta) : \beta = c\alpha\}$.

Conversely, let *D* contain a proper subspace *S* of $\mathbf{R}^2$ other than the coordinate axes. Then $S = \{(\alpha,\beta):\beta = c\alpha\}$ for some $c \neq 0$. Hence, on this subspace, we must have

$$y'' + (-A(x) + cB(x))\alpha y = 0$$

disconjugate on I for every $\alpha \in \mathbf{R}$. Applying lemma 2.1.2 with q(x) = -A(x) + cB(x), we find that A(x) = cB(x) a.e. on I, so that A, B are linearly dependent.

**Corollary 2.1.8 Whenever A, B are linearly independent functions on I, *D* cannot contain any full ray through the origin of parameter space.**

Proof. Note that the assumption of linear independence excludes the possibility that either one of A(x), B(x) vanishes a.e. on I. The result is now a direct consequence of theorem 2.1.7.

**Remark 2.3** Note that if $D$ contains two proper subspaces of $\mathbf{R}^2$ then $D = \mathbf{R}^2$. Thus whenever $D \subsetneq \mathbf{R}^2$, (which is generally the case) $D$ contains at most one full ray through (0,0).

It is important to note that $D$ need not always contain a full ray through (0, 0). In fact, in some cases, $D$ may even be a **bounded** set (see the example below). However note that, for example, $D$ is **always unbounded** whenever A, B are linearly dependent, or if A(x) = 1 a.e. on I and B(x) is arbitrary, as, in this case, $D \supset \{(\alpha, 0):\alpha \geq 0\}$.

**Example 1** We present an example which shows that, under the stated general conditions on A, B, it may occur that $D$ is a bounded set. In order to see this choose A, B as follows.: Let $\varepsilon > 0$, $\eta > 0$ be given numbers and let

$$A(x) = \begin{cases} -\varepsilon, & x \in [0, 4] \\ +\varepsilon, & x \in (4, 8] \\ 0, & x \in (8, \infty). \end{cases}$$

Define B by

$$B(x) = \begin{cases} +\eta, & x \in [0, 1] \cup (2, 3] \cup (4, 5] \cup (6, 7] \\ -\eta, & x \in (1, 2] \cup (3, 4] \cup (5, 6] \cup (7, 8] \\ 0, & x \in (8, \infty). \end{cases}$$

With A, B so defined it is easy to see that (2) is always non-oscillatory on I for any choice of $(\alpha, \beta)$ (since A, B both vanish identically for all x > 8). Thus $N = \mathbf{R}^2$ in this case.

Now it is not difficult, however tedious, to show that there exists $\alpha_0, \beta_0 > 0$ such that $(\alpha, \beta) \notin D$ whenever $|\alpha| \geq \alpha_0$, $|\beta| \geq \beta_0$. That $D \neq \{(0, 0)\}$ can be seen by showing that the rhomb $|\alpha|\,\varepsilon + |\beta|\,\eta = \pi^2/256$ in parameter space, lies in $D$ . Hence $D$ is a nontrivial bounded set. We omit the details.

**Lemma 2.1.9** **Let $q_i$: I → R and in $L_1^{loc}(I)$ for i = 1, 2. Assume that each one of the equations $y'' + q_i(x)y = 0$, i = 1, 2, is disconjugate on I. Then the equation**

$$y'' + ((1-\gamma)q_1(x) + \gamma q_2(x))y = 0 \tag{2.1.9}$$

**is disconjugate on I, for each $\gamma \in [0, 1]$.**

Proof. Let $[a, b] \subset I$ be a compact subinterval. Then the quadratic functional $I(\eta, q_i; a, b)$, $i = 1, 2$, is positive on $A_1(a, b)$, (lemma 2.1.1). On account of the same lemma, it suffices to show that $I(\eta, (1-\gamma)q_1 + \gamma q_2; a, b)$ is positive on $A_1(a, b)$, (since $[a, b]$ is then arbitrary the result will follow). Now if $\gamma \in [0, 1]$ and $\eta \not\equiv 0$ is in $A_1(a, b)$,

$$(1-\gamma)I(\eta, q_1: a, b) + \gamma I(\eta, q_2; a, b) > 0 \tag{2.1.10}$$

However the left side of (2.1.10) is equal to $I(\eta, (1-\gamma)q_1 + \gamma q_2; a, b)$ as a simple calculation will show. Hence the latter is positive definite on $A_1(a, b)$. This is valid for every $[a, b] \subset I$. Therefore the result follows.

**Remark 2.4** The above lemma is similar in spirit to an early result of Adamov [Ad.1, §9] wherein the word "disconjugate" is replaced by "non-oscillatory" in the statement of our lemma. Adamov's result was rediscovered by Moore [Mo.1, lemma 2].

**Lemma 2.1.10** **Let $q_n: I \to R$, $q_n \in L_1^{loc}(I)$ $n = 1, 2, \ldots$ be a sequence of functions such that**

$$y'' + q_n(x)y = 0 \tag{2.1.11}$$

**is disconjugate on I, for each $n = 1, 2, \ldots$**

**If $q_n(x) \to q(x)$ in the $L_1$-sense over each compact subinterval of I, then the limit equation**

$$y'' + q(x)y = 0 \tag{2.1.12}$$

**is also disconjugate on I.**

Proof. Let $x_1, x_2 > 0$ be arbitrary but fixed and consider the interval $[x_1, x_2]$. It is known (see e.g. [Hl.1, Chapter III.1] that the solutions of (2.1.11) - (2.1.12) having the same initial values are "close" in the uniform norm over $[x_1, x_2]$ provided $q_n$ and $q$ are close in $L_1(x_1, x_2)$ which is the case, by assumption.

Now assume, on the contrary, that (2.1.12) is not disconjugate on I. Then it has a solution $y(x) \not\equiv 0$ with two zeros $x_1, x_2$, $x_1 < x_2$, say, in I. This said let $y_n(x)$ be solutions of (1) defined by $y_n(x_1) = 0$, $y'_n(x_1) = y'(x_1)$ $(\neq 0)$. Then for fixed $x > x_1$ and $\varepsilon > 0$, there exists N such that for each $n \geq N$, $\int_{x_1}^{x} |q_n - q|\,dx < \varepsilon$. Hence let $\eta > 0$ be fixed and set $x = x_2 + \eta$. For our $\varepsilon$, there is then a $\delta > 0$ such that

$$\sup_{x \in [x_1, x_2 + \eta]} |y_n(x) - y(x)| < \varepsilon$$

whenever

$$\int_{x_1}^{x_2+\eta} |q_n(x) - q(x)|dx < \delta.$$

Since y(x) must change sign at $x = x_2$, it follows that $y_n(x)$ must also change sign near $x = x_2$ if n is sufficiently large. Thus for such n, (2.1.11) is not disconjugate, which is a contradiction.

**Theorem 2.1.11 In the usual topology of $R^2$, the disconjugacy domain $D$ of (2) is a closed set.**

Proof. Let $(\alpha_0, \beta_0)$ be a limit point of the sequence $(\alpha_n, \beta_n) \in D$, n = 1, 2, ... Then for each $\varepsilon > 0$ there exists an n such that $|\alpha_n - \alpha_0| < \varepsilon$, $|\beta_n - \beta_0| < \varepsilon$ and

$$y'' + (-\alpha_n A(x) + \beta_n B(x))y = 0 \qquad (2.1.13)$$

is not disconjugate. Now let y(x) be any nontrivial solution of (2) for $(\alpha, \beta) = (\alpha_0, \beta_0)$. Either y(x) never vanishes in which case $(\alpha_0, \beta_0) \in D$, (cf. [Ha.4]) or $y(x_0) = 0$ for some $x_0$. In the latter case let $y_n(x)$ be the solution of (2.1.13) which satisfies $y_n(x_0) = 0$, $y'_n(x_0) = y'(x_0)$. Then, by assumption, $y_n(x) \neq 0$ for $x \neq x_0$. However $\{y_n(x)\}$ uniformly approximates y(x) on each interval $[x_0, x_0 + X]$ (by lemma 2.1.10), for $X > 0$ if $\varepsilon$ sufficiently small. Hence y(x) can only change sign at $x = x_0$, and so y(x) $\neq 0$ for $x \neq x_0$ in I. Thus every solution $y(x) \not\equiv 0$ has at most one zero in I. Hence the result follows.

**Theorem 2.1.12. When viewed as a subset of parameter space $R^2$, the disconjugacy domain of (2) is a convex set.**

Proof. We must show that if $(\alpha_i, \beta_i) \in D$, i = 1, 2, then the line segment joining these two points is also in $D$, i.e., that the point $(1-\gamma)(\alpha_1, \beta_1) + \gamma(\alpha_2, \beta_2) \in D$ for each $\alpha \in [0, 1]$. This is equivalent to showing that

$$y'' + ((-(1-\gamma)\alpha_1 - \gamma\alpha_2) A(x) + ((1-\gamma)\beta_1 + \gamma\beta_2) B(x))y = 0$$

is disconjugate on I for each $\gamma \in [0, 1]$. Simplifying and rearranging terms in the potential of the last equation, we may rewrite it in the equivalent form

$$y'' + [(1-\gamma)(-\alpha_1 A(x)+\beta_1 B(x)) + \gamma(-\alpha_2 A(x) + \beta_2 B(x))]y = 0$$

for $\gamma \in [0, 1]$. Since $(\alpha_i, \beta_i) \in D$ for i = 1, 2, lemma 2.1.9 yields the conclusion.

**Example 2** Let $A(x) = B(x) \equiv (x+1)^{-2}$ on $[0, \infty)$. Since A, B are linearly dependent $D$ must contain a full ray through (0, 0). Moreover we know that $D \neq \phi$, $D$ is closed and convex. Note that (2), with the above identifications, becomes an Euler equation, of the form

$$y'' + [(-\alpha + \beta)/(x+1)^2]y = 0$$

for $x \in I$. Thus if $-\alpha + \beta \le \frac{1}{4}$, the equation is disconjugate and so $\{(\alpha, \beta): \beta \le \frac{1}{4} + \alpha\} \subseteqq D$. On the other hand, if $-\alpha + \beta > \frac{1}{4}$, the equation is oscillatory (see e.g. [Sw. 1] for such results). Hence $D = \{(\alpha, \beta) : \beta \le \frac{1}{4} + \alpha\}$. Note that $D$ contains precisely one subspace $S$ of $\mathbf{R}^2$, namely, $S = \{(\alpha, \beta): \alpha = \beta\}$ (cf. theorem 2.1.7).

## 2.2. The Non-Oscillation Domain, $N$

We maintain the notation of §2.1. Recall that (2.1.2) is said to be non-oscillatory at infinity or, more simply, **non-oscillatory** if and only if every non-trivial solution of (2.1.2) has at most finitely many zeros in I.

Since every solution is continuous, and solutions of initial value problems are unique, **(2.1.2) is non-oscillatory if and only if every nontrivial solution changes sign at most finitely many times** (i.e., to each solution $y \not\equiv 0$, there is a finite number of points $\{x_i\}$ such that y(x) actually changes sign around $x_i$, for each i. Note that the $x_i$'s are, of course the zeros of y).

**Remark 2.5** The stated (former) definition of non-oscillation is the classical one. The latter (equivalent) form seems to be the more natural one if one wishes to define a notion of disconjugacy (or non-oscillation) for second order vector differential equations with general matrix coefficients (not necessarily real symmetric), (see §3 for more details).

The non-oscillation domain, $N$, has been defined in §1. Note that $N \neq \phi$, since $D \subseteq N$ and $D \neq \phi$.

We begin this section with examples showing the interplay between $D$ and $N$.

**Example 1.** In contrast with theorem 2.1.5 **one can have $N = \mathbf{R}^2$ without either one of A, B being equal to zero a.e. on I.** A simple way of seeing this is by choosing $A, B \in C_{\circ}[0, \infty) = \{\phi \in C(I) \mid \phi(x)$ vanishes identically outside a closed and bounded subinterval of I$\}$.

For such a choice of A, B it is readily seen that (2) will reduce to y" = 0 for all sufficiently large x, independently of $\alpha$, $\beta$. Hence every solution must have finitely many zeros. So $N = \mathbf{R}^2$ for such potentials.

**Example 2.** We now give an example whereby $\{(0, 0)\} \subsetneq D \subsetneq N \subsetneq \mathbf{R}^2$.

To this end, let A(x) = 1 and B(x) = $(x + 1)^{-2}$ on $[0, \infty)$. Then (2) takes the form

$$y'' + (-\alpha + \beta(x + 1)^{-2})y = 0$$

for $x \in I$. Now it is clear that $D \supset \{(\alpha, \beta) : \alpha \geq 0, \beta \leq \frac{1}{4}\}$ (by a simple application of Sturm's comparison theorem with an Euler equation). Moreover the region $\{(\alpha, \beta) : \alpha < 0\}$ is certainly part of the **oscillation domain** (i.e., the complement in $\mathbf{R}^2$ of the non-oscillation domain) by the Fite-Wintner theorem [Sw.1]. Thus

$$D \neq \{(0, 0)\} \text{ and } N \subsetneq \mathbf{R}^2.$$

**Example 3.** The phenomenon $D = N = \{(0, 0)\}$ may also occur.

It is known that if $q: I \to \mathbf{R}$ is **(Bohr) uniformly almost periodic** of mean-value (M{q}) equal to zero, then (2.1.12) is oscillatory at $+\infty$, (cf., [Mo.2]). In the cited paper Moore showed, in particular, that if q is uniformly almost-periodic then (2.1.12) is either disconjugate on $[0, \infty)$ or oscillatory at $+\infty$, (Actually in [Mo.2] there results are stated for $(-\infty, \infty)$ however they also hold for $[0, \infty)$). Thus **for this class of potentials the notions of disconjugacy and non-oscillation coincide.** Now let A, B each be uniformly almost-periodic functions with M{A} = M{B} = 0. Then $-\alpha A + \beta B$ is also uniformly almost periodic with $M\{-\alpha A + \beta B\} = 0$ for **each** $(\alpha, \beta) \in \mathbf{R}^2 \setminus \{(0, 0)\}$. Hence (2) is oscillatory for each $(\alpha, \beta) \neq (0, 0)$. Thus $D = N = \{(0, 0)\}$.

**Example 4.** In contrast with theorem 2.1.11 we show that, **in general,** $N$ **is not a closed set** by exhibiting an example.

It suffices to find a sequence $\{q_n(x)\}$ such that $q_n(x) \to q(x)$, uniformly on compact subsets of I as $n \to \infty$, and $y'' + q_n(x)y = 0$ is non-oscillatory (for each n) with $y'' + q(x)y = 0$ being oscillatory at $+\infty$. Let $A(x) \equiv x^2$, $B(x) \equiv (x+1)^{-2}$ on $[0, \infty)$, and consider

$$y'' + (-\alpha x^2 + \beta(x+1)^{-2})y = 0 \qquad (2.2.1)$$

on I. Let $(\alpha_n, \beta_n) = (\frac{1}{n}, 1 + \frac{1}{n})$, n = 1, 2, ... . Then (2.2.1) is non-oscillatory at $+\infty$ for each n = 1, 2, ... since $-\alpha_n x^2 + \beta_n(x+1)^{-2} \leq 0$ for each $x \geq X_n \equiv (-1 + \sqrt{1 + 4\sqrt{n+1}})/2$. Hence for each n,

(2.2.1) with $(\alpha, \beta) = (\alpha_n, \beta_n)$ is disconjugate on $[X_n, \infty)$ and hence must be non-oscillatory on $[0, \infty)$. However $-\alpha_n x^2 + \beta_n(x+1)^{-2} \to (x+1)^{-2}$ (uniformly on compact subintervals of $[0, \infty)$), and the limit equation $y'' + (x+1)^{-2}y = 0$ is an oscillatory Euler equation. Thus $(\alpha_n, \beta_n) \in N$ but $\lim (\alpha_n, \beta_n) \notin N$. Hence $N$ cannot be a closed set in this case.

Since $D$ is closed, it follows that $D \neq N$ for the equation (2.2.1).

**Lemma 2.2.1. Let $q_i$: I → R, $q_i \in L_1^{loc}(I)$, for i = 1,2. If each one of the equations $y'' + q_i(x)y = 0$ is non-oscillatory on I then the equation (2.1.9) is also non-oscillatory on I for each $\gamma \in [0, 1]$.**

Proof. The case of continuous $q_i$ may be found in [Mo.1, lemma 2] or [Ad.1]. Note that the former proof uses a variable change, and the Sturm comparison theorem which is well-known to be valid for $q \in L_1^{loc}(I)$, (as Picone's identity holds for such potentials). The proof is therefore similar and so we omit the details.

**Theorem 2.2.2. The non-oscillation domain $N$ of (2) is a convex subset of $R^2$.**

Proof. The proof proceeds along the lines of theorem 2.1.12 except that we now use lemma 2.2.1.

**Remarks on §2.2.** We have seen that, in contrast with D, N is not always closed (example 4) although like D, N is convex (theorem 2.2.2). Furthermore, in contrast with theorem 2.1.5, there may occur $N = \mathbf{R}^2$ without either one of A, B, vanishing a.e. on I, (example 1). In general $D \subsetneq N$ for equation (2) however there exists various different classes of potentials A, B, for which $D = N$ for each member of the class. (see example 2 in §2.1 and example 3 in this section). It follows from the above results that the oscillation domain, $O$, is always connected, $O$ may be an open set (e.g. when $D = N$ as in [Mo.2]) and it may be empty (e.g. when $N = \mathbf{R}^2$ as in example 1). (For comparison purposes it is known [Mo.2, corollary, p. 103] that whenever A(x) = 1, B(x) is uniformly almost-periodic, then $O$ is connected, open and non-empty). The sets $D$, $N$ may be unbounded (e.g. when A, B are linearly dependent, (see theorem 2.1.7)) and $D$ may even be a bounded set (so that $D$ is nonempty, convex and compact), $D \neq \{(0, 0)\}$, (see example 1 of §2.1). Still, $D$ is unbounded in [Mo.2]. It is possible that $N$ may be bounded (as $N = \{(0, 0)\}$ can occur, see example 3). However **it is an open question whether or not $N$ may be bounded if $N \neq \{(0, 0)\}$.** We have also seen that $O = \mathbf{R}^2 \setminus \{(0, 0)\}$ is a possibility (example 3) so that $O$ is as large as possible in such a case. Moreover it is immediate that the boundary of $D$ and of $N$ is a continuous curve (since each one of these sets is convex).

Finally, we wish to point out that consideration of the more general equation

$$(P(x)y')' + (-\alpha A(x) + \beta B(x))y = 0$$

where $P(x) > 0$, $P{:}I \to \mathbf{R}$, $\frac{1}{p} \in L_1^{loc}[0, \infty)$ does not really lead to anything new as it may be transformed into an equation of the form (2) (see [Wl.1]). For this reason we have restricted ourselves to equations of the form (2).

## 2.3 The Location of $D$ and $N$

In this section we give some sufficient conditions of a general nature on A, B which guarantees that the sets $D$, $N$ of the preceding sections are contained in $H^+ \equiv \{(\alpha, \beta) : \alpha > 0\}$, the right-half-plane of parameter space $\mathbf{R}^2$.

First note that whenever $A(x) = 1$ and $B(x)$ is uniformly almost-periodic with $M\{B\} = 0$, then $D \subsetneq \{(\alpha, \beta) : \alpha > 0\} \cup \{0, 0\}$ [Mo.2]. In general, the latter inclusion cannot be expected as it is related to the open problem in §2.4.

**Theorem 2.3.1. Let A(x) satisfy $A(x) \geqq 0$ for large x and**

$$\lim_{x\to\infty} \int_0^x A(s)ds = +\infty. \tag{2.3.1}$$

**Let B(x) satisfy**

$$-\infty < \liminf_{x\to\infty} \int_0^x\int_0^t B(s)ds\, dt < \limsup_{x\to\infty} \int_0^x\int_0^t B(s)ds\, dt < +\infty. \tag{2.3.2}$$

**Then, for equation (2), $D \subseteq N \subsetneq H^+$**

Proof. Let $\alpha < 0$, $\beta \neq 0$. We show that, in this case, (2) is oscillatory. Write

$$v(x) \equiv \int_0^x\int_0^t B(s)ds\, dt, \qquad x \in I. \tag{2.3.3}$$

Now transform (2) according to $y(x) = z(x)\exp(-\beta v(x))$, (as suggested in [Mo.1]). Then $z(x)$ satisfies the equation

$$(\exp(-2\beta v(x))z')' + (-\alpha A(x) + \beta^2 v'(x)^2)\exp(-2\beta v(x))z = 0. \tag{2.3.4}$$

Note that (2) is oscillatory if and only if (2.3.4) is oscillatory. Let $p(x) \equiv \exp(-2\beta v(x))$., $q(x) \equiv (-\alpha A(x) + \beta^2 v'(x)^2)p(x)$. Then (2.3.2) implies that

$$\lim_{x\to+\infty} \int_0^x \frac{ds}{p(s)} = \lim_{x\to+\infty} \int_0^x \exp(2\beta v(s)ds = +\infty$$

Furthermore, for some suitably chosen constant c,

$$\int_0^x q(s)ds > (-\alpha)c \int_0^x A(s)ds, \quad x \in I$$

Hence for $\alpha < 0$, the latter implies that

$$\lim_{x\to\infty} \int_0^x q(s)ds = +\infty.$$

Thus (2.3.4) is oscillatory by Leighton's theorem [Sw.1], (the analog of which is known to be true under "minimal conditions" on A, B as we have here). Therefore for $\alpha < 0$, $\beta \neq 0$, (2) is oscillatory by the Fite-Wintner theorem (i.e. Leighton's theorem with $p(x) \equiv 1$). Hence $N \subseteq H^+$ and the conclusion follows.

**Corollary 2.3.2. (i) For the Mathieu equation**

$$y'' + (-\alpha + \beta\cos 2x)y = 0, \qquad x \in [0, \infty)$$

**we have $D \subseteq N \subseteq H^+$.**

**(ii) Let $B(x) \in C_0^\infty([0, \infty))$.**

**Then, $D \subseteq N \subseteq H^+$ for the equation**

$$y'' + (-\alpha + \beta B(x))y = 0, \qquad x \in [0, \infty).$$

**Remark 2.6** In corollary 2.3.2(ii), as usual, $C_0^\infty([0, \infty))$ stands for the space of infinitely differentiable functions on $[0, \infty)$ which vanish identically outside some compact set contained in $(0, \infty)$. It is now clear that the class of potentials B satisfying (2.3.2) is wide indeed and includes various periodic potentials.

In §2.1 we have seen that $D$ may contain various full rays through the origin of $\mathbf{R}^2$. It should be interesting to determine whether or not $D$ may contain curves of higher (lower) order than one.

For example, a glance at the geometry of the curve $\beta = \alpha^3$ shows that, since $D$ is convex, the line segment joining the points $(\alpha_1, \alpha_1^3)$ and $(\alpha_2, \alpha_2^3)$ on this curve must belong to $D$. However as $\alpha$ varies over $\mathbf{R}$, the convex hull of the set $\{(\alpha, \beta) : \beta = \alpha^3\}$ is, in fact, **all** of $\mathbf{R}^2$. Thus $D = \mathbf{R}^2$ and so the coefficients A, B must vanish a.e. on I, (theorem 2.1.5). From this simple argument there follows that **the equation**

$$y'' + (-\alpha A(x) + \alpha^3 B(x))y = 0, \qquad x \in [0, \infty)$$

**must be non-disconjugate on $[0, \infty)$ for at least one value of $\alpha \in \mathbf{R}$, (if A and/or B are not a.e. zero on I).** A similar argument applies for the general cubic: $\beta = a_1 \alpha^3 + a_2 \alpha^2 + a_3 \alpha + a_4$ where $a_1 \neq 0$, and for the general **odd-degree** polynomial equation (with non-zero leading coefficient).

The case when $\beta = \alpha^2$ (or any even degree polynomial equation) is very different. This is because the convex hull of the set $\{(\alpha, \beta) : \beta = \alpha^2\}$ as $\alpha$ varies over $\mathbf{R}$ is **not** all of $\mathbf{R}^2$, (in contrast with the cubic case mentioned above). The proof of theorem 2.1.5 is practical, as it may be used to derive **necessary** conditions for $D$ to contain the full parabola $\beta = \alpha^2$. For example,

**Theorem 2.3.3. Let $A, B \in L_1^{loc}(I)$ be such that $A^2(x) + B^2(x) > 0$ on a set of positive Lebesgue measure in I.Then a necessary condition for the disconjugacy domain, $D$, of (2) to contain the full parabola $\beta = \alpha^2$, is that $B(x) \leq 0$ a.e. on I.**

Proof. We proceed as in the proof of theorem 2.1.5. Since $D$ contains all of $\{(\alpha, \alpha^2) : \alpha \in \mathbf{R}\}$ it follows that for each interval $[a, b] \subset I$ and each $\eta \neq 0$ in $A_1(a, b)$.

$$I(\eta, -\alpha A + \alpha^2 B; a, b) > 0$$

for **each** $\alpha \in \mathbf{R}$. From this it is readily derived that, for a fixed interval $[a, b] \subset I$ and a given $\eta \not\equiv 0$ in $A_1(a, b)$,

$$\alpha^2 \int_a^b B\eta^2 dt - \alpha \int_a^b A\eta^2 dt < \|\eta'\|_2^2 \tag{2.3.5}$$

for each $\alpha \in \mathbf{R}$.

Since $\eta' \in L_2(a, b)$, it follows that

$$\int_a^b B\eta^2 \leq 0 \tag{2.3.6}$$

for our $\eta$, (else the left-side of (2.3.5) will exceed the right-side for all sufficiently large $\alpha > 0$). Since $\eta \not\equiv 0$ is arbitrary it follows that (2.3.6) holds for each $\eta \not\equiv 0$ in $A_1(a, b)$, (and clearly for $\eta \equiv 0$). Once again, since $[a, b]$ is arbitrary, we find that (2.3.6) holds for each $\eta \in A_1(a, b)$ and each $[a, b] \subset I$. Let $\eta \equiv \phi_\varepsilon$ be the test-function appearing in the proof of theorem 2.1.5. Then, arguing as in the latter, we find

$$\int_a^b B(s)ds \leq 0$$

for each $[a, b] \subset I$. The conclusion now follows.

**Remark 2.7.** Note that the above type of proof may be used to show that $B(x) \le 0$ is, once again, a necessary condition for $D$ to contain the graph of any even-degree polynomial equation with non-zero leading coefficient, (i.e., $\beta = a_0\alpha^{2\eta} + a_1\alpha^{2\eta-1} + \dots + a_{2\eta}$, $a_0 \neq 0$).

Other results in the same vein are the following: *(In the next two theorems let A, B satisfy the hypotheses in theorem 2.3.3).*

**Theorem 2.3.4. A necessary condition that $D$ (of (2)) contains the parabolic segment $\{(\alpha, \beta) : \beta = c\sqrt{\alpha},\ c > 0\}$ for all sufficiently large $\alpha$ (depending on c) is that $A(x) \ge 0$ a.e. on I.**

Proof. This is analogous to the proof of theorem 2.3.3, and so will be omitted.

A special type of converse is

**Theorem 2.3.5. Let $A(x) \ge 0$ a.e. on I, and ess. inf $A(x) > 0$. In addition, let $B \in L_\infty(I)$. If $\alpha \ge \alpha_0$ where**

$$\alpha_0 \equiv c^2 \|B\|_\infty^2/(\text{ess. inf } A(x))^2 \tag{2.3.7}$$

**then $D$ contains the parabolic segment**

$$\{(\alpha, \beta) : \beta = c\sqrt{a}\}$$

**for such $\alpha$.**

**The lower bound $\alpha_0$ in (2.3.7) is sharp when A(x), B(x) are each constant functions a.e. on I.**

Proof. Note that for $\alpha \ge \alpha_0$ we have

$$\sqrt{\alpha}\,(\text{ess. inf } A(x)) - c\,\|B\|_\infty \ge 0$$

and so

$$\sqrt{\alpha}\,A(x) - cB(x) \ge 0 \text{ a.e. on I,}$$

for $\alpha \ge \alpha_0$. Thus

$$\alpha A(x) - (c\sqrt{\alpha})B(x) \ge 0 \tag{2.3.8}$$

a.e. on I for $\alpha \ge \alpha_0$. Now for a given fixed $[a, b] \subset I$ and $\eta \neq 0$ in $A_1$ (a, b) we have

$$\int_a^b \{(\eta'(t))^2 - (-\alpha A(t) + c\sqrt{\alpha}\, B(t))\eta(t)^2\}dt \geq 0 \tag{2.3.9}$$

on account of (2.3.8), with equality holding in (2.3.9) if and only if $\eta \equiv 0$. Hence $I(\eta, -\alpha A + c\sqrt{\alpha}\, B; a, b)$ is positive definite on $A_1$ (a, b) for each $[a, b] \subset I$ and consequently

$$y'' + (-\alpha A(x) + c\sqrt{\alpha}\, B(x))y = 0$$

must be disconjugate on $[0, \infty)$, for $\alpha \geq \alpha_0$, by lemma 2.1.1. Hence $\{(\alpha, c\sqrt{\alpha}) : \alpha \geq \alpha_0\} \subset D$ which is what was required to be shown.

Next let $A(x) \equiv a$, $B(x) \equiv b$ be constant functions a.e. on I. Then it is trivial that $D = \{(\alpha, \beta) : -\alpha a + \beta b \leq 0\}$. Now the boundary of $D$ is the ray $\beta b - \alpha a = 0$, or $\alpha = (\frac{b}{a})\beta$ (since $a > 0$). The point of intersection of the parabolic segment $\beta = c\sqrt{\alpha}$ for $\alpha \geq \alpha_0 = c^2b^2/a^2$ with the boundary of $D$ is given by equating $(\frac{b}{a})\beta = \beta^2/c^2$ which yields $\beta = bc^2/a$, i.e., $\alpha = \alpha_0$. Hence the said parabolic segment originates at the boundary of $D$ for each $c > 0$.

**Remark 2.8.** In theorem 2.3.5 it is possible to show that $D$ contains the parabolic segment $\{(\alpha, c\alpha^{1/r}), \alpha \geq \alpha_0\}$ where $r > 1$, $c > 0$ if

$$\alpha_0 \equiv c^r \, \|B\|_\infty^r / (\text{ess. inf } A(x))^r.$$

The proof is similar.

## 2.4 An open problem.

As was mentioned earlier, Moore [Mo.1] actually obtained the stronger result

$$N \subseteq \{(\alpha, \beta) : \alpha > 0\} \cup \{0, 0\} \tag{2.3.9}$$

in the case when $A(x) \equiv 1$, $B(x)$ is periodic (or Bohr almost-periodic) with mean value equal to zero.

Upon inspection we note that, in order for (2.3.9) to hold, it is necessary that for $\alpha = 0$, and any $\beta \neq 0$, equation (2) must have the property that

$$y'' + \beta B(x)y = 0, \quad x \in [0, \infty)$$

is oscillatory at infinity for each real $\beta \neq 0$. Of course this is not the case in general and so this motivates the following problem -

**Open Problem 1. To characterize, as much as possible, those real valued $r \in L_1^{loc}(I)$ for which the equation**

$$y'' + \lambda r(x)y = 0, \qquad x \in I \text{ or } \mathbb{R}$$

**is oscillatory for each real $\lambda \neq 0$.**

**Remark 2.9** (i) We can easily derive that if r(x) is Bohr almost periodic with mean value equal to zero then r(x) has this property. A function r for which the equation y" + λr(x)y = 0, for x in I or **R**, is oscillatory for each real value of λ not equal to zero, will be termed a **dilation invariant potential relative to oscillation**, for convenience, and abbreviated by "DIPRO". The stated result is a consequence of the fact that the equation

$$y'' + r(x)y = 0 \tag{2.3.10}$$

is oscillatory provided r(x) is a non-trivial Bohr almost periodic function with mean value equal to zero, [Mo.2].

(ii) We note that it was shown by Yelchin [Ye.1] that if r(x) is continuous and periodic and admits a Fourier series whose constant term is zero then (2.3.10) is oscillatory on (-∞, ∞), (cf. [Mo.1]). Since λr(x) has the same property of each real λ ≠ 0, such an r(x) is also a DIPRO (in any case, this result is a consequence of (i) - We mentioned it merely for historical reasons).

(iii) A slightly different class of potentials r which are DIPRO's can be found in Sobol [So.1]. There it is shown that if r(x) is continuous and has a non-constant almost-periodic indefinite integral, then (2.3.10) is oscillatory. Once again λr(x) has the same property for λ ≠ 0 and r is indeed a DIPRO.

(iv) We extend the class of potentials r(x) which solve problem 1 to include functions with may not be (Bohr) almost-periodic.To this end we make use of some results in [At.2]. Therein it is shown, in particular, that the equation

$$y'' - x^a p(x^b) y = 0, \qquad x \in [0, \infty) \tag{2.3.11}$$

where a > b > 0, p is continuous and periodic with mean-value equal to zero, is in the **limit-circle case** at infinity (i.e., all of its solutions are in $L_2(0, \infty)$). It is well-known (and easily derived) that the limit-circle case of an equation of the form (2.3.10) is oscillatory at infinity. Hence (2.3.11) is oscillatory on [0, ∞). Now the function λp(x) where λ ≠ 0 is real, has the same properties as p above and so

$$y'' + \lambda x^a p(x^b) y = 0, \qquad x \in [0, \infty)$$

is oscillatory for each real λ ≠ 0. Hence the potential $r(x) = x^a p(x^b)$ where a > b > 0 and p is as above, is a DIPRO.

However, note that r(x) is not Bohr almost periodic as there always exists a sequence $x_n \to \infty$ with $r(x_n) \to +\infty$ contradicting the uniform boundedness of such r.

A function $r \in L_1^{loc}(I)$ is called an **oscillating potential** if for every $X > 0$ there exists two Lebesgue measurable sets $E^+$, $E^-$, contained in $[X, \infty)$, each of which has finite positive Lebesgue measure and on each of which we have $r(x) > 0$, $r(x) < 0$ respectively.

**Theorem 2.4.1. A necessary condition that the function $r \in L_1^{loc}(I)$ be a DIPRO is that r(x) be an oscillating potential.**

Proof. (i) Assume the contrary. Then there exists $X > 0$ and an interval $[X, \infty)$ in which r(x) is of fixed sign a.e., say, $r(x) > 0$. Since r is a DIPRO, equation (2.1.3) must be oscillatory for every $\lambda \neq 0$ and so, in particular, for $\lambda < 0$. But $\lambda r(x) < 0$ on $[X, \infty)$ and so (2.1.3) is disconjugate on $[X, \infty)$. Moreover $r \in L_1(0, X)$ and so (2.1.3) is non-oscillatory on $[0, X]$. Hence (2.1.3) is non-oscillatory on $[0, \infty)$ which is a contradiction. If $r(x) < 0$ on $[X, \infty)$ let $\lambda > 0$ and apply a similar argument.

**Remark 2.10** The necessary condition of theorem 2.4.1 is not sufficient in general. To see this, let $r(x) = (\sin x)/4(x + 1)^2$ for $x \in I$. Then $r(x) \leq (4(x + 1)^2)^{-1}$ for $x \in [0, \infty)$. However $y'' + [1/4(x+1)^2]y = 0$ is a disconjugate Euler equation. Hence, by the Sturm comparison theorem

$$y'' + \lambda[(\sin x)/4(x + 1)^2]y = 0$$

is also disconjugate on $[0, \infty)$ for $0 < \lambda \leq 1$. Hence this r is an oscillating potential which is not a DIPRO.

We now complement the necessary condition of theorem 2.4.1 with a sufficient condition for a certain class of potentials to be DIPRO's.

**Theorem 2.4.2 Let $r \in L_1^{loc}(I)$ and $\int_{x_0}^{x_1} r(s)ds \neq 0$ for some $x_0 < x_1$ in I. If**

$$-\infty < \liminf_{x \to +\infty} v(x) \leq \limsup_{x \to +\infty} v(x) < \infty \tag{2.3.12}$$

**where**

$$v(x) = \int_{x_0}^{x} \int_{x_0}^{t} r(ds)ds\, dt, \tag{2.3.13}$$

**and, in addition,**

$$\lim_{x\to+\infty} \int_{x_0}^{x} \{ \int_{x_0}^{t} r(s)ds \}^2 dt = + \infty, \tag{2.3.14}$$

**the r(x) is a DIPRO.**

Proof. We refer to the proof of theorem 2.3.1. Set $\alpha = 0$ therein and now define $v(x)$ as in (2.3.13). Once again the transformation $y(x) = z(x) \exp\{-\lambda v(x)\}$ will transform (2.1.3) into an equation of the form

$$(p(x)z')' + q(x)z = 0, \quad x \in I$$

where $p(x) = \exp(-2\lambda v(x))$, $q(x) = \lambda^2 v'(x)^2 p(x)$. Now (2.3.12) implies that

$$\lim_{x\to\infty} \int_0^x \frac{ds}{p(s)} = + \infty$$

for any $\lambda \neq 0$.

Moreover, for $\lambda \neq 0$, note that $p(x)$ is bounded away from zero for large x (by (2.3.12)). Moreover $v'(x) \not\equiv 0$ by hypothesis. Hence

$$\lim_{x\to+\infty} \int_{x_0}^{x} q(s)ds = \lim_{x\to+\infty} \int_{x_0}^{x} \lambda^2 v'(s)^2 p(s)ds \geqq \lambda^2(\text{const}) \int_{x_0}^{\infty} \{ \int_{x_0}^{x} r(s)ds\}^2 dx$$

by (2.3.14). The result is now a consequence of Leighton's theorem [Sw.11].

**Remark 2.11** (i) Note that the collection of functions satisfying (2.3.12) and (2.3.14) is non-empty as $r(x) = \cos x$, ($x_0 = 0$) is such a function.

(ii) Condition (2.3.12) by itself implies that r(x) cannot remain of fixed sign a.e. on every interval of the form $[X, \infty)$ where $X > 0$. Hence r must be an oscillating potential.

(iii) The novelty of theorem 2.4.2 lies in the fact that there are no assumptions regarding the mean value of r. In fact, this quantity need not even exist. However, if the oscillating potential in question **does** admit a mean value more can be said -

(iv) For a result related to theorem 2.4.2 see [Mi.4, proposition 3].

**Theorem 2.4.3.** **Let $r \in L_1^{loc}(I)$ satisfy $\int_0^{x_1} r(s)ds \neq 0$ for some $x_1 > 0$ and assume that (2.3.12) holds. If r(x) admits a mean-value over $[0, \infty)$, i.e.,**

$$\lim_{x\to+\infty} \left\{\frac{1}{x}\int_0^x r(s)ds\right\} = A \tag{2.3.15}$$

**exists and is finite, then A = 0 (≡ M{r}).**

Proof. For if $A > 0$, let $\varepsilon = \frac{A}{2}$. Then there exists a $x_1 > 0$ such that for each $x \geq X_1$,

$$\int_0^x r(s)ds \geq \frac{Ax}{2}$$

Hence

$$\int_{x_0}^x \int_0^t r(s)ds\, dt \geq \frac{A}{2}(x^2 - X_1^2)$$

for $x \geq X_1$. Hence

$$\liminf_{x\to+\infty} \int_{X_1}^x \int_0^t r(s)ds\, dt = +\infty$$

and so (for $x_0 = 0$)

$$\liminf_{x\to+\infty} v(x) = +\infty$$

which contradicts (2.3.12).

If $A < 0$, let $\varepsilon = -\frac{A}{2}$. Then there exists $X_2 > 0$ such that for each $x \geq X_2$,

$$\int_0^x r(s)ds \leq \frac{Ax}{2}.$$

Proceeding as above we find

$$\liminf_{x\to+\infty} \int_{X_2}^x \int_0^t r(s)ds\, dt = -\infty.$$

which gives the contradiction

$$\liminf_{x\to+\infty} v(x) = -\infty.$$

**Remark 2.12.** (i) In the above proof we only used the weaker form of (2.3.12), namely, that

$$-\infty < \liminf_{x\to+\infty} v(x) < +\infty.$$

(ii) One may speculate at this point that if r is an oscillating potential with $M\{r\} = 0$, then r is a DIPRO. However such is **not** generally the case for we may set $r(x) = (\sin x)/(x+1)^3$, $x \in [0, \infty)$: Then r is such a potential. However

$$y'' + \lambda(\sin x)/(x+1)^3 y = 0$$

is non-oscillating for $-\infty < \lambda < +\infty$ (by comparison with an Euler equation).

(iii) It now follows from the above that those r which satisfy the hypotheses of theorem 2.4.2 are in particular oscillating potentials with mean value equal to zero (if such a mean-value exists).

## 2.5 Applications to the Sturm-Liouville equation and its extensions.

Let $A, B \in L_1^{loc}(I)$, as usual, and consider the weighted Sturm-Liouville equation

$$y'' + (\lambda B(x) - A(x))y = 0 \tag{2.5.1}$$

where $x \in I$, $(=[0, \infty))$, and $\lambda \in \mathbf{R}$ is a parameter. Note that (2.5.1) is a special case of (2) with the identifications $\alpha = 1$, $\beta = \lambda$.

Now we turn our attention to the collection of those $\lambda \in \mathbf{R}$ for which (2.5.1) is non-oscillatory /oscillatory. The collection of such $\lambda$ is a vertical line $L$ through $\alpha = 1$ in parameter space $\mathbf{R}^2$. Hence we will investigate the number of possible ways in which $L$ may intersect $N$ (or $D$).

**Remark 2.13** (i) Using the technique developed in the proof of lemma 2.1.2., we can show:

If equation (2.5.1) is disconjugate on $[0, \infty)$ for every real $\lambda$, the $B(x) = 0$ a.e. on I (and $y'' = A(x)y$ is disconjugate on I). Therefore it follows that **if $B \neq 0$ on a set of positive measure then there exists at least one value of $\lambda$ for which (2.5.1) is not disconjugate on I.**

(ii) The results of §2.3 imply the following: Let A, B be real (non-trivial) Bohr almost-periodic functions with $M\{A\} = M\{B\} = 0$. Then (2.5.1) is oscillatory at infinity for every real value of $\lambda$. From this we find

**Theorem 2.5.1.** **Let A, B each be real ($\not\equiv 0$) Bohr almost-periodic functions with $M\{A\} = M\{B\} = 0$.Then any boundary problem associated with (2.5.1) on I (or $(-\infty, \infty)$) will have the property that its real eigenvalues have only oscillatory eigenfunctions. In particular, such a problem has no real ground state (i.e., a real eigenfunction which has no zeros in the interior of the interval under consideration).**

**Theorem 2.5.2.** **Let A, B $\in L_1^{loc}(I)$ . Then precisely one of the following five (5) cases occurs for each equation of the form (2.5.1):**

**(a) It is oscillatory for every real $\lambda$.**

**(b) It is oscillatory for every real value of $\lambda$ except at some unique point $\lambda = \lambda_0$.**

**(c) There exists a finite interval $(\lambda_1, \lambda_2)$ in R such that (2.5.1) is oscillatory for $\lambda \in (-\infty, \lambda_1) \cup (\lambda_2, \infty)$ and non-oscillatory for $\lambda \in (\lambda_1, \lambda_2)$.**

**(d) There exists a point $\lambda_3 \in$ R such that (2.5.1) is oscillatory (resp. non-oscillatory) on $(-\infty, \lambda_3)$ and non-oscillatory (resp. oscillatory) for $\lambda \in (\lambda_3, \infty)$.**

**(e) It is non-oscillatory for every real $\lambda$.**

Proof. Let $L = \{(1, \lambda) : \lambda \in \mathbf{R}\}$ be the ray mentioned at the beginning of this section. Since $L$ is a ray and $N$ is convex the claims (a) - (e) are consequences of the geometrical nature of the intersection of $L$ with $N$ . Recall that $N$ is a convex set in general position in $\mathbf{R}^2$ (but, as always, containing (0, 0)). Enumerating the possibilities is now an easy matter; There are only five distinct ways in which $L$ may intersect $N$ : (a) $L \cap N = \phi$; (b) $L \cap N$ = {a single point} (c) $L \cap N$ = {a finite line segment}; (d) $L \cap N$ = {a "half-ray"}; (e) $L \cap N$ = {a full ray} = $L$ . Each of the possibilities (a) - (e) listed here corresponds to the stated claims (a) - (e) (respectively) in the theorem, as is not difficult to see.

**Example 1. We show here that, in fact, each one of the possibilities (a) - (e) stated in theorem 2.5.2 may occur.**

On (a): See theorem 2.5.1.

On (b): See theorem 2.5.1 but now set A $\equiv$ 0 there. Then $y'' + \lambda B(x)y = 0$ will be oscillatory for each real $\lambda \neq 0$ by [Mo.2], and so $L \cap N = \{(1, 0)\}$ in this case.

On (c): Let B be continuous and purely periodic of period one and M{B} = 0, and $A(x) \equiv 1$, $x \in (-\infty, \infty)$. In this case $N$ (=$D$ ) is contained in $\{(\alpha, \beta) : \alpha > 0\}$ and meets the $\beta$-axis only at (0, 0). Since the boundary of $N$ is a continuous curve (as it is convex) and it is formed by the union of the graphs of two unbounded monotone functions, $L \cap N$ must be a finite line segment. (For details on the said structure of $N$ see [Mo.1])

On (d): Let $B(x) \equiv (x+1)^{-2}$ on $[0, \infty)$ and set $A(x) \equiv 0$ on $[0, \infty)$. Then for $\lambda \in (-\infty, 1/4)$ (2.5.1) is non-oscillatory (in fact, disconjugate) while for $\lambda \in (1/4, +\infty)$, (2.5.1) is oscillatory at $\infty$ as the equation is of Euler-type. Replacing $B(x)$ by $-B(x)$ and 1/4 by -1/4 interchanges the words "non-oscillatory" and "oscillatory".

On (e): Let $A, B \in C_0^\infty\ [0, \infty)$. Then $N$ (for equation (2)) is all of $\mathbf{R}^2$ and so (2.5.1) is non-oscillatory for every real $\lambda$, (see example 1, §2.2.).

**Theorem 2.5.3. Let $A \not\equiv 0$ be a Bohr almost periodic function with mean-value $M\{A\}$. Then there exists a value $\lambda^*$ of $\lambda$ such that the equation**

$$y'' + (\lambda - A(x))y = 0, \qquad x \in \mathbf{R}, \tag{2.5.2}$$

**is disconjugate for $\lambda \in (-\infty, \lambda^*)$ and oscillatory (at $\pm\infty$) for $\lambda \in (\lambda^*, \infty)$. Furthermore**

$$\inf_{x \in \mathbf{R}} A(x) \leq \lambda^* \leq M\{A\} \tag{2.5.3}$$

**The estimates in (2.5.3) are precise.**

Proof. In this case of equation (2.5.1) with $B(x) \equiv 1$, it is known [Ha.2] that precisely one of the following three cases must occur: (i). (2.5.2) is oscillatory for all $\lambda \in \mathbf{R}$; (ii). It is oscillatory for $\lambda \in (-\infty, \lambda^*)$ and non-oscillatory for $\lambda \in (\lambda^*, \infty)$ for some $\lambda^* \in \mathbf{R}$; (iii). It is non-oscillatory for every $\lambda \in \mathbf{R}$.

Now let $M\{A\} = \lambda_0$. Then the function $\lambda_0 - A$ is Bohr almost periodic and $M\{\lambda_0 - A\} = 0$. Hence (2.5.2) with $\lambda = \lambda_0$ is oscillatory by [Mo.2]. Moreover since A is almost periodic it is bounded (uniformly) on $\mathbf{R}$. Hence for $\lambda < \inf \{A(x): x \in \mathbf{R}\}$, $\lambda - A(x) < 0$ on $(-\infty, \infty)$ and so (2.5.2) is disconjugate for such $\lambda$. Hence there must exist $\lambda^* \in \mathbf{R}$ such that (2.5.2) is disconjugate for $\lambda \in (-\infty, \lambda^*)$ and oscillatory for $\lambda \in (\lambda^*, \infty)$. The bounds is (2.5.3) are attained for $A \equiv$ constant $\neq 0$.

Furthermore (2.5.3) must hold in this case by previous considerations.

This result now enables us to formulate various results regarding the spectrum of a self-adjoint problem associated with (2.5.2).

**Corollary 2.5.4. The part of the spectrum of any self-adjoint extension of the operator**

$$\frac{-d^2}{dx^2} + A(x), \qquad x \in (-\infty, \infty) \tag{2.5.4}$$

**where A is Bohr almost-periodic, is finite on $(-\infty, \lambda^*)$ and infinite on $(-\infty, \lambda^* + \varepsilon)$ for each $\varepsilon > 0$.**

Proof. This follows from theorem 2.5.3 and a classical result of Hartman [Ha.5] which relates the oscillatory/non-oscillatory behavior of solutions of (2.5.2) for a value of $\lambda$ to the number of points in the spectrum of a self-adjoint extension of (2.5.4) around $\lambda$.

It is known [Mo.2, lemma 1] that (1.2) is oscillatory on $(-\infty, \infty)$ if and only if it is oscillatory on $[0, \infty)$ or $(-\infty, 0]$. Hence we may use oscillation theory on a half-axis to our advantage...

Let $f:[0, \infty) \to \mathbf{R}$ be continuous. We define [KZ1] extended numbers $\overline{\alpha}$, $\underline{\alpha}$ by

$$\overline{\alpha}(f) = \inf \{\mu \in \mathbf{R} : \int_0^\infty [f(t) - \mu]_+^2 \, dt < \infty\}$$

$$\underline{\alpha}(f) = \sup \{\mu \in \mathbf{R} : \int_0^\infty [f(t) - \mu]_-^2 \, dt < \infty\}$$

Here $f_\pm(x) = \max\{\pm f(x), 0\}$ as usual.

**Lemma 2.5.5 [KZ1, theorem 2]. Let $v(x) = \int_0^x V(s)ds$. If either**

**(i) $-\infty < \underline{\alpha}(v) < \overline{\alpha}(v)$ or**

**(ii) $\underline{\alpha}(v) < \overline{\alpha}(v) < +\infty$**

**holds, then**

$$y'' + V(x)y = 0 \tag{2.5.5}$$

**is oscillatory at $+\infty$.**

**Lemma 2.5.6 [Mo.3, theorem 9]**

**Let $V:[0, \infty) \to \mathbf{R}$ be continuous. If $V(x) \le M^2 < \infty$ for $x \in [0, \infty)$ and**

$$\limsup_{x \to \infty} \int_0^x V(s)ds = +\infty \tag{2.5.6}$$

**then (2.5.5) is oscillatory at $+\infty$.**

**Corollary 2.5.7** **The corresponding result on $(-\infty, 0]$ is the following: If $V(x) \le M^2 < +\infty$ for $x \in (-\infty, 0]$ and**

$$\liminf_{x \to -\infty} \int_0^x V(s)ds = -\infty. \tag{2.5.7}$$

**then (2.5.5) is oscillatory at $-\infty$.**

**Lemma 2.5.8** **[Wi.1].** **Let $V: [0, \infty) \to R$, $V \in L_1^{loc}(I)$, (R). If**

$$\lim_{x\to\infty} \int_t^x V(s)ds$$

**exists (for all sufficiently large t) and**

$$-\frac{3}{4} \le t \int_t^\infty V(s)ds \le \frac{1}{4},$$

**then, (2.5.5) is oscillatory at $+\infty$.**

**Note:** The continuity restriction on V in [Wi.1] is unnecessary and only $L_1^{loc}(I)$ behavior is really required. (see, e.g., [Mo.3, theorem 6] wherein the proof holds under this more general assumption) In the sequel, a function f is said to be in the class H(V), where V is Bohr almost periodic, provided f is the uniform limit [on $(-\infty, \infty)$] of a sequence of translates of V (viz, f(x) is the uniform limit of a sequence $V(x+t_n)$ for some sequence $t_n$ of real numbers). This set is usually called the **hull** of V.

**Lemma 2.5.9** **[Fa.1, p. 48]**

**Let V be a almost-periodic function with $M\{V(x)\} = 0$ and assume that V has an unbounded (above and below) indefinite integral.**

**Then there exists, in H(V), at least two functions $V^+$, $V^-$ such that**

$$\int_0^x V^+(s)ds \ge 0, \quad \int_0^x V^-(s)ds \le 0,$$

**for each x, $-\infty < x < +\infty$.**

**Lemma 2.5.10** **Let $v \ne 0$ be almost-periodic. Then $v \notin L_2(R)$.**

Proof. Write $g = v^2$. Then $g(x) \geq 0$ is also a.p. and in fact, $M\{g(x)\} > 0$ (by Bohr's uniqueness theorem, [Bo.1, p. 63]. Now for $\varepsilon > 0$ choose $T_0(\varepsilon) > 0$ so large that

$$\varepsilon > \frac{1}{T}\int_0^T g(s)ds - M\{g(x)\} > -\varepsilon$$

for each $T \geq T_0$. Then, choosing $\varepsilon = M\{g(x)\} / 2$ we find that for $T \geq T_1$,

$$\int_0^T g(s)ds > (M\{g(x)\} / 2)T$$

and so $g \notin L(0, \infty)$. The result follows.

**Lemma 2.5.11** **Let v be almost-periodic. Then $|v|$, $v_+ = \max\{v, 0\}$, $v_- = \max\{-v, 0\}$ are also almost-periodic and so $M\{v_\pm^2(x)\}$ exists and is finite.**

Proof. The first statement is clear. Moreover

$$v_+(x) = (v(x) + |v(x)|)/2, \qquad v_-(x) = (|v(x)| - v(x))/2$$

and so, since sums (differences) of a.p. functions are a.p. [Bo.1, p. 36, theorem 3], the conclusion follows.

**Theorem 2.5.12** **Let $V \neq 0$ be almost-periodic function.**

**Then a necessary and sufficient condition for $-y'' + \lambda V(x)y = 0$ to be oscillatory at $+\infty$ for every real $\lambda \neq 0$ is that $M\{V(x)\} = 0$.**

**Corollary 2.5.13** **The following statements are equivalent:**

**(i)** **The equation**

$$-y'' + \lambda V(x)y = 0, \qquad x \in R \tag{2.5.8}$$

**is oscillatory at $\pm\infty$ for every real $\lambda \neq 0$.**

**(ii)** **There exists finite numbers $\lambda^+ > 0$, $\lambda^- < 0$, such that (2.5.8) is oscillatory at $\pm\infty$ for every $\lambda \in (\lambda^-, \lambda^+)$, $\lambda \neq 0$.**

**(iii)** **$M\{V(x)\} = 0$.**

Proof. That (i) <=> (iii) and (i) => (ii) is clear. That (ii) => (i) follows from the convexity of the disconjugacy domain [Mo.2].

**Note.** The case V purely periodic with $M\{V(x)\} = 0$ of the theorem 3.1 can be found in Stanek [Sk.1]. However this case actually follows directly from [Mo.2, theorem 2 and theorem 6].

**Proof of Theorem 2.5.12** (Sufficiency) We divide this part of the proof into two principal cases:

1. The case when V has an unbounded integral.

2. The case when $v(x) := \int_0^x V(s)ds$ is uniformly bounded on **R**.

Case 1. Assume $\limsup_{x\to+\infty} \int_0^x V(s)ds = +\infty$. Since V is a.p. it is uniformly bounded and so lemma 2.5.6 implies that (2.5.5) is oscillatory at $+\infty$ and so also at $-\infty$.

On the other hand if (2.5.6) fails to hold, then, by lemma 2.5.9. there exists a function $V^*(x)$ in H(V) for which

$$\int_0^x V^*(s)ds \leq 0$$

for $-\infty < x < \infty$ and, consequently, for such a function $V^*$ one must have

$$\liminf_{x\to-\infty} \int_0^x V^*(s)ds = -\infty$$

Hence corollary 2.5.7 implies that the equation

$$y'' + V^*(x)y = 0, \qquad x \in \mathbf{R} \tag{2.5.9}$$

is oscillatory at $-\infty$. But $V^*(x) = \lim_{i\to\infty} V(x+\tau_i)$, uniformly of **R**, for some sequence $\{\tau_i\} \subset \mathbf{R}$. Now

if

$$y'' + V(x+\tau_i)y = 0 \tag{2.5.10}$$

is not oscillatory at $+\infty$ then it must be disconjugate on **R**, [Mo.2, lemma 1]. However the uniform limit of disconjugate equations of the form (2.5.10) must also be a disconjugate equation. Hence (2.5.9) is disconjugate, which is a contradiction. Thus (2.5.10) is oscillatory at $+\infty$ and also at $-\infty$ and hence the same must be true for (2.5.5). It follows that if $M\{V(x)\} = 0$ and V has an unbounded integral then (2.5.5) is oscillatory at $\pm\infty$.

Case 2. In this case v must be a.p. [Bo.1, p. 58]. Hence $v_+$, $v_-$, $v_+^2$, $v_-^2$ are all a.p. by lemma 2.5.11 and so the corresponding mean values $M\{v_\pm^2(x)\}$ all exist and are finite. There are now four possibilities:

(1) $M\{v_+^2(x)\} > 0$ ; (2) $M\{v_+^2(x)\} = 0$ ; (3) $M\{v_-^2(x)\} > 0$ ;

(4) $M\{v_-^2(x)\} = 0$.

Note that (2) implies (3) and that (4) implies (1) (by the Bohr uniqueness theorem [Bo.1, p. 63] and because $v(x) \not\equiv 0$ for all x). Hence, once again, there are two cases i.e.,

(i) $M\{v_-^2(x)\} > 0$

(ii) $M\{v_+^2(x)\} > 0$.

On (i): Since $M\{v_-^2(x)\} > 0$, $v_- \not\equiv 0$ so lemma 2.5.10 implies that $v_- \notin L_2(\mathbf{R})$. Thus $\underline{\alpha}(v) \leq 0$.

Moreover $\underline{\alpha} \leq \overline{\alpha}$, [KZ1, proposition 1]. Since we also have

$$\overline{\alpha}(v) = \sup\{\mu \in \mathbf{R}: \int_0^\infty [v(s) - \mu]_+^2 \, ds = +\infty\}$$

we see that

$$\int_0^\infty [v(s) - \mu]_+^2 \, ds \equiv 0$$

for $\mu \geq \sup \{|v(x)|: x \in \mathbf{R}\}$. Hence $\overline{\alpha} < \infty$, i.e., we get

$$\underline{\alpha}(v) \leq \overline{\alpha}(v) < \infty \tag{2.5.11}$$

On (ii): $M\{v_+^2(x)\} > 0$ implies once again that $v_+ \notin L_2(\mathbf{R})$ and so $\overline{\alpha}(v) \geq 0$. Once again for $\mu \leq -\sup \{|v(x)| : x \in \mathbf{R}\}$ we have

$$\int_0^\infty [v(s) - \mu]_-^2 ds < \infty$$

i.e., $\underline{\alpha}(v) \geq -\sup \{|v(x)|: x \in \mathbf{R}\}$ and so

$$-\infty < \underline{\alpha}(v) \leq \overline{\alpha}(v) \tag{2.5.12}$$

Thus, in either case (i) or (ii), we have (2.5.11) or (2.5.12) holding and so there remains to show that $\underline{\alpha} < \overline{\alpha}$. (as lemma 2.5.5 will yield that (2.5.5) is oscillatory at $+\infty$ and so also at $-\infty$).

Finally, let $\mu < \sup\{v(x) : x \in \mathbf{R}\}$. Then the continuity of v implies that $(v-\mu)_+ \not\equiv 0$ on **R**. However, since $(v-\mu)_+$ is a.p., lemma 2.5.10 implies that $(v-\mu)_+ \notin L_2(\mathbf{R})$, i.e.,

$$\overline{\alpha}(v) \geq \sup\{v(x) : x \in \mathbf{R}\} \tag{2.5.13}$$

Similarly if $\mu > \inf\{v(x) : x \in \mathbf{R}\}$, then $(v-\mu)_- \not\equiv 0$ on **R** and so, once again, $(v-\mu)_- \notin L_2(\mathbf{R})$.

Thus

$$\underline{\alpha}(v) \leq \inf\{v(x) : x \in \mathbf{R}\} \tag{2.5.14}$$

Combining (2.5.13-14) we obtain

$$\underline{\alpha}(v) \leq \inf\{v(x) : x \in \mathbf{R}\} < \sup\{v(x) : x \in \mathbf{R}\} \leq \overline{\alpha}(v)$$

(as v(x) cannot be identically constant without V vanishing identically on **R**). Thus $\underline{\alpha}(v) < \overline{\alpha}(v)$. In this first part of the theorem we have shown that if $M\{V(x)\} = 0$, then (2.5.5) is oscillatory at $\pm\infty$. (This was also shown in [Mo.2, theorem 2] however their proof for the case when v is unbounded is incomplete as the stated conclusion [Mo.2, p. 103] does <u>not</u> follow from previous considerations).

Hence from this it follows that whenever $M\{V(x)\} = 0$, the equation (2.5.8) is oscillatory at $\pm\infty$ for every real $\lambda \neq 0$.

(Necessity). We will show that whenever $M\{V(x)\} \neq 0$, there exists a value of $\lambda \in \mathbf{R}$, $\lambda \neq 0$, for which (2.5.5) is disconjugate on **R**. It will follow from this that (2.5.8) will be oscillatory for every real $\lambda \neq 0$, only if $M\{V(x)\} = 0$.

To this end let $M\{V(x)\} = m \neq 0$ and consider the single differential equation in the two real parameters $\mu$, $\nu$,

$$y'' + (-\nu + \mu V(x))y = 0 \tag{2.5.15}$$

on $[0, \infty)$. (Note that non-oscillation on $[0, \infty)$ implies disconjugacy on $[0, \infty)$ and so on $(-\infty, \infty)$ by results in [Mo.2]). Then (2.5.15) may be rewritten as

$$y'' + (-\alpha + \beta V^*(x))y = 0 \tag{2.5.16}$$

where $\beta = \mu$ and $\alpha = \nu - m\mu$, and $M\{V^*(x)\} = 0$.

(Let $\alpha > 0$, $\beta \neq 0$). We now make the transformation $y = z\exp(-x\sqrt{\alpha})$ and $t = (1/2\sqrt{\alpha})\exp(2\sqrt{\alpha}\,x)$ in (2.5.16). This leads us to an equation

$$z'' + \beta e^{-\sqrt{\alpha}\,x} V(x) z = 0$$

and $f(t) = \beta V(x) \exp(-4\sqrt{\alpha}\, x)$. The x-interval $(-\infty, \infty)$ goes into the half-axis, $[0, \infty)$.

Now

$$t \int_t^\infty f(s)ds = \frac{\beta e^{2\sqrt{\alpha x}}}{2\sqrt{\alpha}} \int_x^\infty V(s) \exp(-2\sqrt{\alpha}\, s)ds,$$

(note that both these integrals converge since V is bounded),

$$= \frac{\beta}{2\sqrt{\alpha}} \int_0^\infty e^{-2\sqrt{\alpha\tau}} V(x+\tau)d\tau.$$

$$= \beta \int_0^\infty e^{-2\sqrt{\alpha}\,\tau} \int_0^\tau V(x+s)ds\, d\tau$$

$$= \beta \int_0^\infty \tau e^{-2\sqrt{\alpha}\,\tau} [\frac{1}{\tau} \int_0^\tau V(x+s)ds] d\tau$$

$$= \beta \int_0^\infty \tau e^{-2\sqrt{\alpha}\,\tau} [\frac{1}{\tau} \int_x^{x+\tau} V(s)ds]\, d\tau \tag{2.5.17}$$

Since $M\{V(x)\} = 0$ and V is a.p. then for every $\varepsilon > 0$, there exists $\tau_0(\varepsilon) > 0$ for which

$$\sup_{x \in R} \left| \frac{1}{\tau} \int_x^{x+\tau} V(s)ds \right| \le \in \tag{2.5.18}$$

for $\tau \ge \tau_0$, [Bo.1, p. 44].

Thus let $T > 0$ and rewrite (2.5.17) as an integral over $[0, T]$ plus an integral over $[T, \infty)$. Then

$$\left| \int_0^T \tau e^{-2\tau\sqrt{\alpha}} [\frac{1}{\tau} \int_x^{x+\tau} V(s)ds] d\tau \right| \le M \int_0^T \tau e^{2\sqrt{\alpha\tau}} d\tau \tag{2.5.19}$$

as it is certainly the case that the integral appearing in the square parentheses is bounded, by $M = M(T)$ say, as it is a continuous function of $\tau \in [0, T]$. (Note that $\sup\{M(t): T \ge 0\} < \infty$ on account of (2.5.18).

Moreover, since $M\{V(x)\} = 0$ we have

$$\int_{\tau}^{\infty} \tau e^{-2\sqrt{\alpha}\,\tau}\Big[\frac{1}{\tau}\int_{x}^{x+\tau} V(s)ds\Big]\,d\tau \mid \le \sup_{\tau\in[T,\infty)} \Big|\frac{1}{\tau}\int_{x}^{x+\tau} V(s)ds\Big| . \int_{T}^{\infty} \tau e^{-2\sqrt{\alpha}\,\tau}d\tau. \tag{2.5.20}$$

$$\le \varepsilon(T) \int_{T}^{\infty} \tau e^{-2\sqrt{\alpha}\,\tau}d\tau \tag{2.5.21}$$

Combining the estimates (2.5.19), (2.5.21) and writing $K = 2\sqrt{\alpha}\,T$ we obtain

$$\Big|\int_{t}^{\infty} f(s)ds\Big| \le \frac{M|\beta|}{4\alpha}[1-(K+1)e^{-K}] + \frac{\varepsilon|\beta|}{4\alpha}(K+1)e^{-K} \le \frac{|\beta|}{4\alpha}\{\frac{M}{2}K^2 + \varepsilon(T)\} \tag{2.5.22}$$

We may now let $T \to \infty$ in such a way that $T = 0(\alpha^{-1/2})$ as $\alpha \to 0^+$. Then, the uniformity of the mean value (2.5.18), will imply that $\varepsilon(T) \to 0$ uniformly in x (see (2.5.20)). Moreover we will also have $K \to 0$ (as $\alpha \to 0^+$).

Hence if $\alpha > 0$ is sufficiently small we see that

$$\frac{|\beta|}{4\alpha}\{\frac{M}{2}K^2 + \varepsilon(T)\} \le \frac{1}{4}, \tag{2.5.23}$$

i.e., if

$$|\beta| \le \alpha\psi(\alpha) \tag{2.5.24}$$

where $\psi(\alpha) = \{MK^2/2 + \varepsilon(T)\}^{-1}$ for an appropriately large T which we then fix, then (2.5.16) will be non-oscillatory (and so disconjugate) on $[0, \infty)$ on account of lemma 2.5.8, i.e., (2.5.16) will be disconjugate on $(-\infty, \infty)$. It follows from (2.5.24) that the disconjugacy domain just touches the $\beta$-axis at the origin, and at (0, 0) we have a vertical tangent!

We now return to (2.5.16). Assume $m > 0$. We set $\nu = 0$ in (2.5.15), i.e., $\alpha = -m\mu = -m\beta$ in (2.5.16). Then from the preceding discussion it follows that the line $\alpha + m\beta = 0$ must intersect the disconjugacy domain of (2.5.16) for some $\alpha > 0$ and some range of negative $\beta$'s, say, $0 > \beta \ge \beta_0$. (Similarly if $m < 0$, we may find such a range of positive $\beta$'s, $0 < \beta \le \beta_1$). In either case there exists $\mu \neq 0$ for which (2.5.15) (with $\nu = 0$) is disconjugate on **R**. This completes the proof of the necessity and of the theorem.

**Remark** The proof of the necessity shows that the disconjugacy domain of an equation (2.5.16) with $V^*$ a.p. and $M\{V^*(x)\} = 0$ has a (boundary with a) vertical tangent at (0, 0) and lies completely in the right half-plane $\{\alpha > 0\} \cup \{(0, 0)\}$. This extends a corresponding result of Markus and Moore [Mo.2; p. 106, theorem 6] wherein it is further assumed that v(x) (defined earlier) is also a.p.

**Further results**

The technique of proof used in the preceding section allows us to state further results ... Thus, for example, it follows from case 2 of the sufficiency part of the theorem that there holds

**Theorem 2.5.14** **Let $0 \not\equiv V(x) = \sum_{n=1}^{N} (A_n \cos \lambda_n x + B_n \sin \lambda_n x)$, $\lambda_n$, $A_n$, $B_n \in \mathbf{R}$, have $M\{V(x)\} = 0$. Then (2.5.5) is oscillatory at $+\infty$.**

**Remark** Indeed, we must have $\lambda_n \neq 0$, for each n, by hypothesis. Thus we may assume $\lambda_n > 0$ for all n, n=1, ..., N. If we write $\lambda = \min \{\lambda_1, \lambda_2, ..., \lambda_N\}$ (all $\lambda_i > 0$) then a simple computation show that, for example, if $t_i \uparrow \infty$ is any sequence then

$$\int_0^{t_i} (A_n \cos \lambda_n x) dx \geq -\frac{2\pi}{\lambda_n} |A_n|, \quad \text{each } i$$

since the period of $\cos \lambda_n x$ is $2\pi/\lambda_n$. Hence

$$v(t_i) = \int_0^{t_i} V(s) ds \geq -2\pi \sum_{n=1}^{N} \frac{1}{\lambda_n} (|A_n| + |B_n|)$$

and, in fact,

$$\left| \int_0^{t_i} V(s) ds \right| \leq 2\pi \sum_{n=1}^{N} \frac{1}{\lambda} (|A_n| + |B_n|)$$

$$\leq \frac{2\pi}{\lambda} \left( \sup_{x \in \mathbf{R}} |V(x)| \right)$$

Hence for a real trigonometric polynomial V with $M\{V(x)\} = 0$ we have shown that,

$$|v(x)| \leq \frac{CM}{\lambda}, \quad -\infty < x < +\infty$$

where $C > 0$ is an absolute constant and $M = \sup \{ |V(x)| : x \in \mathbf{R} \}$, a well-known inequality [Bo.2]. This implies that v is a.p. and allows us to use case 2 of the said theorem.

We refer the reader to [Si.2] for further results involving almost-periodic potentials. For results relating to "non-definite" Sturm equations (2.5.1) see [Mi.2], [Mi.3] and the references therein.

## 2.6 Applications to equations of the form $y'' + (\lambda^2 p(x) + \lambda r(x) - q(x))y = 0.$

Assume that some information has been obtained for the structure of $N$ for an equation

$$y'' + (-\alpha A(x) + \beta B(x))y = 0 \tag{2.6.1}$$

where $x \in I$ and, as usual, $A, B \in L_1^{loc}(I)$.

Consider the parabola $P\ (\alpha_0, \gamma, \delta) \equiv P$ defined by the locus of points $(\alpha, \beta) \in \mathbf{R}^2$ for which $\beta = \gamma(\alpha - \alpha_0)^2 + \delta$. Then $(\alpha, \beta) \in P \cap N$ if and only if the equation

$$y'' + (\alpha^2 p(x) + \alpha r(x) - q(x))y = 0 \tag{2.6.2}$$

is non-oscillatory for the said $\alpha$. Here, $p(x) \equiv \gamma B(x)$; $r(x) \equiv -(A(x) + 2\alpha_0\gamma B(x))$; $q(x) \equiv -(\gamma\alpha_0^2 + \delta)B(x)$, (Note that p, q are linearly dependent). Thus the large-scale structure of the set of all $\alpha$ for which (2.6.2) is non-oscillatory can be determined from (2.6.1) by enumerating the number of possible ways in which $P$ can intersect $N$ .

A complete analog of theorem 2.5.2 for the **general** equation (2.6.2) (i.e. with p, q, all r unrestricted as to sign) is, at the present time, unknown. However if $p(x) \geq 0$ there is such a analog.

**Theorem 2.6.1 Let p, q, r $\in L_1^{loc}(I)$ and $p(x) \geq 0$ a.e. on I. Then precisely one of each of the five possibilities (a) - (e) mentioned in theorem 2.5.2 must occur for every equation of the form**

$$y'' + (\lambda^2 p(x) + \lambda r(x) - q(x))y = 0 \tag{2.6.3}$$

Proof. Conditions (a) - (e) are clear from earlier considerations (as $p(x) = 0$ a.e. on I is not excluded). We will now show that there is at most **one** (maximal) interval of non-oscillation under the stated assumption on p.

To this end, let $\lambda < \mu$ be two points in $\mathbf{R}$ for which each one of the equations (2.6.3) and (2.6.3) with $\lambda$ replaced by $\mu$, is non-oscillatory.

We know that (lemma 2.2.1) the equation $y'' + Q(x, \gamma) = 0$ will be non-oscillatory for each $\gamma \in [0, 1]$ where $Q(\bullet,\gamma) \equiv \gamma(\lambda^2 p + \lambda r - q) + (1-\gamma)(\mu^2 p + \mu r - q)$. Now Q may be rewritten in the form

$$Q(x,\gamma) = (\gamma\lambda^2 + (1-\gamma)\mu^2)p(x) + (\gamma\lambda + (1-\gamma)\mu)r(x) - q(x). \tag{2.6.4}$$

Write $\lambda^* \equiv \gamma\lambda + (1-\gamma)\mu$. Then $\lambda < \lambda^* < \mu$ since $\gamma \in [0, 1]$. Moreover a simple calculation shows that

$$\lambda^{*2} \leq \gamma\lambda^2 + (1-\gamma)\mu^2 \tag{2.6.5}$$

for each $\gamma \in [0, 1]$. Thus $Q(x, \gamma) \geq \lambda^{*2}p(x) + \lambda^* r(x) - q(x)$, a.e. on $[0, \infty)$. The Sturm comparison theorem now implies that

$$y'' + (\lambda^{*2}p(x) + \lambda^{*}r(x) - q(x))y = 0$$

is non-oscillatory, for each $\gamma \in [0, 1]$. However $\lambda^*$ runs through every value in $[\lambda, \mu]$ as $\gamma$ varies from 0 to 1. Hence the whole interval $[\lambda, \mu]$ must be an interval of non-oscillation. Clearly there cannot be two disjoint intervals of non-oscillation by this argument. Hence there can be at most one such (maximal) interval and this completes the proof.

If $p(x)$ is unrestricted as to its sign (or even if $p(x) < 0$ a.e. on I) a classification of the type suggested by theorem 2.6.1 is difficult to formulate. The following partial result can nevertheless be shown:

**Theorem 2.6.2** **Let p, q, r $\in L_1^{loc}(I)$, p(x) unrestricted as to its sign but $r(x) \geq 0$ a.e. on I, (resp. $r(x) \leq 0$ a.e. on I).**

**Then there is at most one (maximal) interval of non-oscillation contained in $(-\infty, 0]$ (resp. $[0, \infty)$).**

Proof. Let $r(x) \geq 0$ a.e. on I (the proof of the case $r(x) \leq 0$ is obtained by replacing r by -r and $\lambda$ by $-\lambda$ in the ensuing discussion). Define $\lambda, \mu$ as in the proof of theorem 2.6.1 except that now $\lambda, \mu \in (-\infty, 0]$. Then, as in the said proof, we have $\lambda_*^2 = \gamma\lambda^2 + (1-\gamma)\mu^2 \geq \lambda^{*2}$ from which we find $\lambda_* \geq |\lambda^*|$ or $-\lambda_* \leq \lambda^*$. Hence

$$\lambda_*^2 p(x) + \lambda^* r(x) - q(x) \geq \lambda_*^2 p(x) - \lambda_* r(x) - q(x) = (-\lambda_*)^2 p(x) + (-\lambda_*) r(x) - q(x).$$

However $-\lambda_* \in [\lambda, \mu]$ (as $\lambda^2 \geq \gamma\lambda^2 + (1-\gamma)\mu^2 \geq \mu^2$ and $\lambda < \mu \leq 0$). Moreover the proof of theorem 2.6.1 gives that

$$y'' + (\lambda_*^2 p(x) + \lambda^* r(x) - q(x))y = 0$$

is non-oscillatory. Hence the Sturm comparison theorem implies that

$$y'' + (\lambda_*^2 p(x) - \lambda_* r(x) - q(x))y = 0$$

is also non-oscillatory. But $-\lambda_*$ takes on every value between $\lambda$ and $\mu$. Hence $[\lambda, \mu]$ is an interval of non-oscillation and so there can be at most one such (maximal) interval of non-oscillation).

**Example 1.** In this example we show that if the sign condition on p(x) is dropped (e.g. in theorem 2.6.2) then there exists a parabola $P(\alpha_0, \gamma, \delta)$ which intersects $N$ for an equation (2.6.1) in **four** points. It will follow that **there may exist up to five (pairwise disjoint, maximal) intervals of oscillation/ non-oscillation.**

We refer the reader to [Mo.1] for background information regarding the results mentioned below. Consider the equation

$$y'' + (\beta B(x) - \alpha)y = 0 \qquad (2.6.6)$$

where B(x) is some fixed, non-constant, continuous, periodic function with $M\{B\} = 0$, and $\alpha$, $\beta$ are real parameters. Then the non-oscillation domain $N$ of (2.6.6) resembles a parabola whose interior contains the semi-axis $\alpha > 0$ and whose vertex is at (0, 0). Writing

$\beta B(x) - \alpha \equiv \lambda^2 p(x) + \lambda r(x) - q(x)$, we set $\beta = \gamma(\alpha - \alpha_0)^2 + \delta$ where $\alpha_0$, $\gamma$, $\delta$ will be chosen below. Then for $\lambda = \alpha$ we find

$$\begin{aligned} \beta B(x) - \alpha &= (\gamma(\alpha-\alpha_0)^2 + \delta)B(x) - \alpha \\ &= (\gamma B(x))\alpha^2 - (1+2\alpha_0\gamma B(x))\alpha + (\gamma\alpha_0^2 + \delta)B(x) \\ &= p(x)\lambda^2 + r(x)\lambda - q(x) \end{aligned}$$

where the identifications are clear. We now refer to the geometry of the intersecting domains $P$ and $N$. Fix some $\alpha_0 > 0$. Then there must exist $\delta < 0$ (and a $\gamma > 0$) such that the vertex of $P$ $(\alpha_0, \gamma, \delta)$ will lie "below" $N$. In this case the parabola $P$ will intersect $N$ in four points (for $\gamma$ sufficiently large). The resulting choice of $\alpha_0$, $\gamma$, $\delta$ then generates an equation of the form (2.6.3) for which there are precisely five intervals of oscillation/non-oscillation. More precisely there are three intervals of oscillation and two intervals of non-oscillation in this case.

Of course p(x) changes sign here so that theorem 2.6.1 cannot apply. Furthermore since $\alpha_0 > 0$, $\gamma > 0$ we cannot have $r(x) \geq 0$ on $[0, \infty)$ (since this would imply that $B(x) < 0$ on I and this is impossible since B has mean value equal to zero). However, it is conceivable that $r(x) \leq 0$ on I. In this case one of the intervals of non-oscillation referred to above must lie on $[0, \infty)$ by theorem 2.6.2.

**Open problem 2. Let p, q, r $\in L_1^{loc}(I)$ and p, q, r all unrestricted as to their sign.**

**What is the largest number of (maximal) intervals of oscillation-non-oscillation that (2.6.3) may have on the real $\lambda$-axis?**

**Note** It is not even known whether or not there may be infinitely many such intervals.

**Remark 2.14** An inspection of the proof of theorem 2.6.1 shows that lemma 2.2.1 is crucial. Now the stronger version of lemma 2.2.1 is lemma 2.1.9. Hence re-interpreting the proof of theorem 2.6.1 with the word "non-oscillatory" replaced by "disconjugate" there follows that there can be an at most one (maximal) interval at each point of which (2.6.3) is disconjugate. From this we find

**Theorem 2.6.3** **Let p, q, r $\in L_1^{loc}(I)$ (or $L_1^{loc}(R)$). Assume further that $p(x) \geq 0$ a.e. on I (or R). Then the collection of all those $\lambda \in R$ for which (2.6.3) admits a positive solution in $(0, \infty)$ (or $(-\infty, \infty)$) constitutes a unique (maximal) interval of the real $\lambda$-axis.**

Proof. This is clear from remark 2.14 above as (2.6.3) has a positive solution on $(0, \infty)$ (or $(-\infty, \infty)$) if and only if (2.6.3) is disconjugate on the respective intervals.

The next result yields some information on the length of the intervals of non-oscillation of (2.6.3).

**Theorem 2.6.4.** **Let p, q, r $\in L_1^{loc}(I)$ and $p(x) \geq 0$ a.e. on I. Assume further that the equation**

$$y'' + (\lambda_0^2 p(x) + \lambda_0 r(x) - q(x))y = 0 \tag{2.6.7}$$

**is non-oscillatory and that**

$$\operatorname*{ess.\,inf}_{x \in [0,\infty)} \left\{ \frac{r(x)}{p(x)} + 2\lambda_o \right\} \geq 0. \tag{2.6.8}$$

**Then there is a left-neighborhood $[\lambda_0 - \varepsilon, \lambda_0]$ of $\lambda_0$ (which may possibly reduce to $\lambda_0$ itself) in which (2.6.3) will be non-oscillatory as long as**

$$0 \leq \varepsilon \leq \operatorname*{ess.\,inf}_{x \in I} \left( \frac{r(x)}{p(x)} + 2\lambda_o \right\}. \tag{2.6.9}$$

Proof. The assumption (2.6.9) implies that

$$\varepsilon^2 p(x) - 2\lambda_0 p(x)\varepsilon \leq \varepsilon r(x) \tag{2.6.10}$$

a.e. on I since $p(x) \geq 0$ a.e. by hypothesis. Thus completing the square in the left-side of (2.6.10) we find

$$p(x)\,(\lambda_0 - \varepsilon)^2 \leq \lambda_0^2 p(x) + \varepsilon r(x),$$

i.e.,

$$p(x)(\lambda_0 - \varepsilon)^2 + r(x)(\lambda_0 - \varepsilon) - q(x) \leq \lambda_0^2 p(x) + \lambda_0 r(x) - q(x) \tag{2.6.11}$$

a.e. on I. But (2.6.7) is non-oscillatory by hypothesis. Hence the Sturm comparison theorem implies that (2.6.3) is non-oscillatory for each $\varepsilon$ satisfying (2.6.9) (on account of (2.6.11)).

**Remark 2.15** (1) We note that for some r, p, $\lambda_0$, (2.6.8) actually yields the **maximum** length of an interval of non-oscillation.

To see this let $p(x) = x + 1$, $r(x) = -x$ and $q(x) \equiv 1$ on $[0, \infty)$. Then the equation

$$y'' + (\lambda^2(x+1) - \lambda x - 1)y = 0 \qquad (2.6.12)$$

is disconjugate for $\lambda \in [0, 1]$ and oscillatory otherwise.

Thus if we let $\lambda_0 = 1$ in theorem 2.6.4, (2.6.9) will imply that (2.6.12) will be non-oscillatory on $[1-\varepsilon, 1]$ for each $\varepsilon \in [0, 1]$, i.e., (2.6.12) will be non-oscillatory for $\lambda \in [0, 1]$ which is precisely the **whole** interval of non-oscillation.

(ii) Finally a result analogous to theorem 2.6.4 may be formulated if $\lambda_0$ is a point of oscillation for (2.6.7). Thus it can be shown that **if (2.6.7) is oscillatory and**

$$\operatorname*{ess.\,sup}_{x \in I} \left\{\frac{r(x)}{p(x)} + 2\lambda_0\right\} \geq 0,$$

**then (2.6.3) will be oscillatory on $[\lambda_0 + \varepsilon, \infty)$ provided**

$$\varepsilon \geq \operatorname*{ess\,sup}_{x \in I} \left\{\frac{r(x)}{p(x)} + 2\lambda_0\right\}.$$

The proof of this result is similar and so is omitted.

## 2.7 On the equivalence of $D$ and $N$.

The results in [Mo.1, Mo.2] motivate the next question: **For which classes of potentials A, B does equation (2) have the property that $D = N$ ?**

That $D = N$ may occur was seen in [Mo.2] where the special case $A \equiv 1$, B almost-periodic was considered. However there are classes of non-oscillating potentials for which $D = N$ also.

**Example 1.** Let $A(x) = B(x) \equiv (x+1)^{-2}$ for $x \in [0, \infty)$. Then, for equation (2), $D = N$ (see example 2, §2.1).

**Example 2.** Let $A(x) \geq 0$ a.e. on $[0, \infty)$ and $A(x) \equiv B(x)$ where $A \in L_1^{loc}(I)$. Assume further that

$$\lim_{x \to \infty} \int_0^x A(s)ds = +\infty. \qquad (2.7.1)$$

Then the equation

$$y'' + (-\alpha+\beta)\, A(x)y = 0$$

is oscillatory whenever $\beta > \alpha$, on account of (2.7.1) and the Fite-Wintner theorem [Sw.1] or [Mi.2,; theorem 2.2.3]. Moreover if $\beta \leq \alpha$, we have $(\beta - \alpha)\, A(x) \leq 0$ a.e. on I and so (2.7.1) is disconjugate.

Hence $O = \{(\alpha, \beta) : \beta > \alpha\}$ and so $D = N\,( = \{(\alpha, \beta) : \beta \le \alpha\})$.

Now any one of a multitude (c.f., [Ba.1], [Wi .1] of oscillation criteria for second order equations with positive coefficients may be used instead of (2.7.1) to obtain still wider classes of potentials for which $D = N$.

Because of the relation of this question to §2.3 (and [Mo.2]) we will now restrict ourselves to oscillatory potentials, more general than those encountered in [Mo.2].

In order to complement the said results of Markus and Moore **we will now restrict ourselves to the interval** $(-\infty, \infty) \equiv \mathbf{R}$ (instead of $I \equiv [0, \infty)$).

Now there are many generalizations of the notion of almost-periodicity as introduced by Bohr (e.g., [Bo.1]) which waive the restrictive requirement that the function in question be continuous on **R**, (see [Be.1]). One of the first generalizations of Bohr almost-periodicity is due to Stepanov [St.1] - This is the extension we wish to consider below.

In the sequel $q \in L_1^{loc}(\mathbf{R})$, $q$: $\mathbf{R} \to \mathbf{R}$. We say that $q \in S_L$, **the class of Stepanov almost-periodic functions** if there exists a real number $L > 0$ such that for each $\varepsilon > 0$, there exists a relatively dense set of Stepanov translation numbers $\tau_q(\varepsilon)$, i.e., numbers such that

$$\sup_{x \in \mathbf{R}} \left\{ \frac{1}{L} \int_x^{x+L} |q(s + \tau_q(\varepsilon)) - q(s)|\,ds \right\} < \varepsilon$$

(see [Be.1] or [Bo.1, Appendix I] for terminology and the basic theory of the class $S_L$).

This definition of Stepanov almost periodic functions may be found in [Be.1, p. 77] and is the classical one. It is to be noted however that the Stepanov class alluded to above may be found by taking the closure relative to the metric $D_S$ defined by

$$D_S[f(x), g(x)] = \sup_{-\infty<x<+\infty} L^{-1} \int_x^{x+L} |f(s) - g(s)|\,ds,$$

of the class of all finite trigonometric polynomials. The notion of a **Weyl almost periodic function** is defined analogously, the only difference being in the definition of the metric. Thus, the Weyl metric is given by

$$D_W[f(x), g(x)] = \lim_{L\to\infty} \sup_{-\infty<x<+\infty} L^{-1} \int_x^{x+L} |f(s) - g(s)|\,ds$$

where the limit may be shown to exist [Be.1, p. 82]. The completion of the class of all finite trigonometric polynomials relative to this metric gives the space of the Weyl a.p. functions.

The generalization of almost periodic functions undertaken by Besicovitch [Be.1, p. 77] is as follows. The Besicovitch metric is defined by

$$D_B[f(x), g(x)] = \limsup_{T\to\infty} \frac{1}{2T} \int_{-T}^{+T} |f(s) - g(s)|\,ds$$

and the space of all **Besicovitch almost periodic functions** is obtained by completing the space of all finite trigonometric polynomials relative to the Besicovitch metric. For each one of these three notions of generalized almost periodic functions, the **mean value** exists i.e.,

$$M\{f(x)\} = \lim_{T\to\infty} \frac{1}{2T} \int_{-T}^{+T} f(s)\,ds$$

exists for each such generalized a.p. function f [Be.1, p. 93, lemma 4].

It is interesting to note that the necessity part of theorem 2.5.12 merely requires the existence of the mean value, $M\{f(x)\}$, albeit it needs to be uniform with respect to translation of f, viz.,

$$\lim_{T\to\infty} \frac{1}{2T} \int_{-T}^{+T} f(s+a)\,ds = M\{f(x)\}$$

uniformly with respect to a, thus allowing for the extension of the said necessity to these cases cf. theorem 2.8.8-9 below and [HM.1], [HM.2].

Our first result extends lemma 1 in [Mo.2] from the Bohr almost-periodic case to the Stepanov almost-periodic case.

**Theorem 2.7.1. Let $q \in \mathcal{S}_L > 0$. If the equation**

$$y'' + q(x)y = 0, \qquad x \in R \tag{2.7.2}$$

**is not disconjugate on $(-\infty, \infty)$ then it is, in fact, oscillatory at both $\pm\infty$.**

Proof. Let $y \neq 0$ satisfy (2.7.2) and be such that for some $a < b$, $y(a) = y(b) = 0$.

Let $\varepsilon = |n|^{-1}$, $|n| = 1, 2, \ldots$. Then there exists (by definition) arbitrarily large positive and arbitrarily small negative $\{\tau_n\}_{n=-\infty}^{\infty}$ (arranged in increasing order) so that, for each n,

$$\sup_{x\in R} \{\frac{1}{L} \int_x^{x+L} |q(s+\tau_n) - q(s)|\,ds\} < {}^{1}/|n|.$$

Hence

$$\frac{1}{L}\int_{a+iL}^{a+(i+1)L} |q(s+\tau_n) - q(s)|\,ds < {}^{1}/|n| \tag{2.7.3}$$

for each i; i = 0, 1, 2, ... . Since L > 0 is fixed, let $m \in \mathbf{N}$ be chosen so that mL > b-a. Fix such an m. Then (2.7.3) holds for i = 0, 1, ..., m-1, and so

$$\frac{1}{L}\int_{a}^{a+mL} |q(s+\tau_n) - q(s)|\,ds < m/|n|$$

holds for each n; n = ± 1, 2, ... . Now there exists some $\eta > 0$, which we now fix, such that

$$\frac{1}{L}\int_{a}^{b+\eta} |q(s+\tau_n) - q(s)|\,ds < m/|n| \tag{2.7.4}$$

for each n. It follows from (2.7.4) that given $v > 0$ there exists N > 0 such that

$$\int_{a}^{b+\eta} |q(s+\tau_n) - q(s)|\,ds < v \tag{2.7.5}$$

provided $|n| \geq N$. Now consider the equations

$$z_n'' + q(x+\tau_n)z_n = 0, \qquad x \in \mathbf{R}$$

and let $z_n(x)$ be the solution corresponding to $z_n(a) = y(a) = 0$, $z'_n(a) = y'(a)$ ($\neq 0$). Then $z_n$ is defined and continuous on $[a, b+\eta]$ and since $q(x+\tau_n)$ approximates $q(x)$ in the $L_1$ - sense over $[a, b+\eta]$ (by (2.7.5)) it follows by the continuous dependence of solutions [H1.1] that there holds

$$\sup_{x\in[a,\, b+\eta]} |z_n(x) - y(x)| < \varepsilon_n$$

where $\varepsilon_n \to 0$ as $|n| \to \infty$. Since y(b) = 0, there exists a n such that $z_n(x)$ will vanish in some neighborhood $(b - \delta, b + \delta)$ of b where $\delta$ may be chosen $\leq \min\{b - a, \eta\}$. Hence $z_n(x)$ will vanish at two points for all $|n| \geq N^*$. But $z_n(x) \equiv y_n(x+\tau_n)$ where $y_n \not\equiv 0$ satisfies (2.7.2) and $y_n(a+\tau_n) = 0$, $y_n'(a+\tau_n) = y'(a)$. Hence $y_n(x)$ vanishes at $x = a + \tau_n$ and near $x = b + \tau_n$ for each n. The Sturm separation theorem now implies that every solution of (2.7.2) must vanish between $a + \tau_n$ and $b + \tau_n + \delta$. Thus (2.7.2) is oscillatory at $\pm\infty$.

**Remark 2.** It is not known at this time whether the class $S_L$ may be replaced by the somewhat "larger" class of Weyl almost-periodic functions, obtained by letting $L \to \infty$ in the definition of the Stepanov translation numbers, or by the more general class of Besicovitch almost-periodic functions.

**Corollary 2.7.2.** **Let A, B $\in S_L$ for some fixed L > 0. Then $D = N$ for equation (2).**

Proof. Since A, B $\in S_L$ and $\alpha, \beta \in \mathbf{R}$, the function $\beta B - \alpha A \in S_L$, [Be.1]. Hence either $(\alpha, \beta) \in D$ or $(\alpha, \beta) \in O$ , by theorem 2.7.1. The result is now clear.

## 2.8 Disconjugacy and Stepanov almost-periodic potentials.

We say that a **sequence $\{g_n(x)\}$** of functions in $L_1^{loc}(\mathbf{R})$ **converges in the Stepanov metric to a function g** provided that, for some L > 0,

$$\sup_{x \in \mathbf{R}} \{\frac{1}{L} \int_x^{x+L} |g_n(s) - g(s)| ds\} \to 0$$

as $n \to \infty$. When this occurs, we say that **g is the Stepanov limit of the sequence $g_n$** and write this symbolically as

$$g(x) = \underset{n \to \infty}{S\text{-}\lim}\ g_n(x).$$

**Theorem 2.8.1.** **Let $q \in S_L$, L > 0 and assume that**

$$y'' + q(x)y = 0, \qquad x \in \mathbf{R} \tag{2.8.1}$$

**is disconjugate on R. Then**

$$y'' + q^*(x)y = 0, \qquad x \in \mathbf{R} \tag{2.8.2}$$

**is also disconjugate on** R **every function $q^* \in SH(q)$, the Stepanov closed hull of q.**

Proof. $q^* \in SH(q)$ if and only if

$$q^*(x) = \underset{n \to \infty}{S\text{-}\lim}\ q(x + \tau_n)$$

for some sequence $\tau_n \subset \mathbf{R}$. The hypothesis that (2.8.1) is disconjugate on $\mathbf{R}$ implies that

$$y'' + q(x + \tau_n)y = 0 \tag{2.8.3}$$

is also disconjugate on $\mathbf{R}$, for each n. Thus (2.8.3) must be disconjugate on (x, x + L), for each $x \in \mathbf{R}$. By definition,

$$\sup_{x \in \mathbf{R}} \{\frac{1}{L} \int_x^{x+L} |q^*(s) - q(s + \tau_n)| ds\} \to 0$$

as $n \to \infty$. Hence $q(x + \tau_n)$ approximates $q^*(x)$ in the $L_1$-sense over (x, x + L), for each x. We now proceed along the lines of theorem 2.7.1.

Assume, if possible, that (2.8.2) is not disconjugate on **R**. Then there exists a < b and a solution $y \not\equiv 0$ of (2.8.2) such that y(a) = y(b) = 0. Let $m \in N$ be chosen so that mL > b - a. As before we can show that, for some $\eta > 0$,

$$\int_a^{b+\eta} |q^*(s) - q(s + \tau_n)|ds \to 0$$

as $n \to \infty$, so that $q^*(x)$ is approximated arbitrarily closely in $L_1(a, b + \eta)$. Continuous dependence of the solutions on the coefficients in this sense then yields the result, (cf. theorem 2.7.1).

**Lemma 2.8.2 [Ha.1] Let $q: [0, \infty) \to R$ be continuous. If there exists an $\varepsilon > 0$ such that**

$$\int_t^{s+t} q(u)du$$

**is half-bounded for $t \in [0, \infty)$ and $0 \le s \le \varepsilon$, then a necessary condition that (2.8.1) be non-oscillatory is that either**

$$\lim_{t\to\infty} \frac{1}{T} \int_0^T\int_0^s q(u)du\, ds = C$$

**exists and is finite, or that**

$$\lim_{T\to\infty} \int_0^T q(s)ds = -\infty, \tag{2.8.4}$$

We will use lemma 2.8.2 in order to obtain a necessary condition for the non-oscillation of solutions of (2.8.1) in the case when q is a Stepanov almost-periodic potential.

**Theorem 2.8.3. Let $q \in \mathcal{S}_L$, L > 0, be continuous on $(-\infty, \infty)$ and assume that (2.8.1) is non-oscillatory on R. Write $Q(x) = \int_0^x q(s)ds$. If Q(x) fails to have a (finite) mean-value (over $[0, \infty)$) then**

$$\lim_{T\to\infty} \int_0^T q(s)ds = -\infty$$

Proof. The hypotheses imply that (2.8.1) is disconjugate on $[0, \infty)$, (theorem 2.7.1). Since $q \in S_L$, it is bounded in the Stepanov metric [Be.1; p.83, 2° lemma], i.e., there exists a $M > 0$ such that

$$\sup_{x \in \mathbf{R}} \left[ \frac{1}{L} \int_x^{x+L} |q(s)|\,ds \right\} \le M < \infty.$$

In particular we must have

$$\int_x^{x+L} |q(s)|\,ds \le ML, \quad x \in [0, \infty).$$

From this there follows that

$$ML \ge \int_x^{x+\xi} q(s)ds \ge -ML, \quad x \in [0, \infty)$$

for each $\xi \in [0, L]$. The conclusion follows by a direct application of lemma 2.8.2.

The next results are in the same spirit as that of theorem 2.5.12. Following Stepanov we define, for each $h \neq 0$, $h \in \mathbf{R}$, a function $F_h : \mathbf{R} \to \mathbf{R}$ defined by

$$F_h(x) \equiv \frac{1}{h} \int_x^{x+h} f(y)dy \tag{2.8.5}$$

where $f \in S_L$, $L > 0$, is the space of Stepanov a.p. functions.

**Lemma 2.8.4. Let f, $F_h$ be as above. The $F_h$ is Bohr almost-periodic (for each $h \neq 0$) and**

$$\mathbf{M\{F_L(x)\} = M\{f(x)\}} \tag{2.8.6}$$

**Proof.** The first part is due to Stepanov [St.1]. In order to prove the second part we write, (for $T \neq 0$),

$$M_T\{F(x)\} \equiv \frac{1}{T} \int_0^T F_h(x)dx. \tag{2.8.7}$$

Then, for $h > 0$,

$$M_T\{F_h(x)\} = \frac{1}{hT} \int\int_R f(y)dy\,dx$$

where $R = \{(x,y): x \le y \le x + h;\ 0 \le x \le T\}$. Using Fubini's theorem it is readily shown that

$$M_T\{F_h(x)\} = \frac{1}{Th} \sum_{i=1}^{3} \iint_{S_i} f\,dx\,dy$$

where

$$S_1 = \{(x,y)\colon 0 \le x \le y,\ 0 \le y \le h\}.$$

$$S_2 = \{(x,y)\colon y - h \le x \le y,\ h \le y \le T\}$$

$$S_3 = \{(x,y)\colon y - h \le x \le T,\ T \le y \le T + h\}.$$

Thus we may write

$$M_T\{F_h(x)\} = \sum_{i=1}^{3} \{\frac{1}{Th} \iint_{S_i} f\,dx\,dy\}$$

$$\equiv I_1 + I_2 + I_3. \tag{2.8.8}$$

It is easy to see that

$$I_1 = \frac{1}{Th} \int_0^h yf(y)\,dy \tag{2.8.9}$$

for each $h > 0$. Furthermore,

$$I_2 = \frac{1}{T} \int_h^T f(y)dy$$

$$= M_T\{f(x)\} - \frac{1}{T} \int_0^h f(y)dy$$

$$= M_T\{f(x)\} - \frac{1}{hT} \int_0^T hf(y)dy \tag{2.8.10}$$

for each $h > 0$. Next, the change of variable $t = y - T$ shows that

$$I_3 = \frac{1}{hT} \int_0^h (h-t)\, f(t+T)dt \tag{2.8.11}$$

Combining (2.8.9-10-11) we find, via (2.8.8), that

$$M_T\{F_h(x)\} = \frac{1}{Th} \int_0^h (y-h)f(y)dy + M_T\{f(x)\} + \frac{1}{hT} \int_0^h (h-y)f(y+T)dy$$

$$= M_T\{f(x)\} + \frac{1}{Th}\int_0^h (y-h)\{f(y)-f(y+T)\}dy$$

$$\equiv M_T\{f(x)\} + I_4 \qquad (2.8.12)$$

Now

$$|I_4| \le \frac{1}{T}\int_0^h |f(y)-f(y+T)|dy = \frac{h}{T}\{\frac{1}{h}\int_0^h |f(y)-f(y+T)|dy\}$$

Thus, setting h = L, and using (2.8.12) we find that

$$|M_T\{F_L(x)\} - M_T\{f(x)\}| \le \frac{1}{T}\{\frac{1}{L}\int_0^L |f(y)-f(y+T)|dy\} \qquad (2.8.13)$$

Let $\varepsilon > 0$ be given. Since $f \in S_L$ there exists $l(\varepsilon,h) > 0$ such that each interval $(x, x+l)$ contains a point $\tau$ for which

$$\sup_{x\in\mathbf{R}} \{\frac{1}{L}\int_x^{x+L} |f(y) - f(y+\tau)|dy\} \le \varepsilon$$

If we choose our intervals $(x, x+l)$ of the form $(i, i+L)$, $i = 0,1,2,...$ then within each such interval there is a $\tau_i \in \mathbf{R}$ for which

$$\frac{1}{L}\int_0^L |f(y) - f(y+\tau_i)|dy \le \varepsilon$$

The sequence $\{\tau_i\}$ is clearly unbounded. Use of (2.8.13) with $T = \tau_i$ shows that

$$|M_{\tau_i}\{F_L(x)\} - M_{\tau_i}\{f(x)\}| \le \frac{L}{\tau_i}\varepsilon < \varepsilon. \qquad (2.8.14)$$

if i is sufficiently large. Since it is known that $M_T\{F_L(x)\}$ converges to the mean-value of $F_L(x)$ as $T \to +\infty$ and the same is true of $M_T\{f(x)\}$, as $f \in S_L$, we deduce from (2.8.14) that there holds (2.8.6).

**Corollary 2.8.5.** **Let $f \in S_L$ for some $L > 0$. If $f(x) \ge 0$ a.e. and $M\{f(x)\} = 0$ then $f(x) = 0$ a.e. on R.**

**Proof.** By lemma 2.8.4, $M\{f(x)\} = M\{F_L(x)\} = 0$. But $F_L$ is a Bohr a.p. function and $F_L(x) \ge 0$. Since $M\{F_L(x)\} = 0$ it follows that $F_L(x) \equiv 0$ on $\mathbf{R}$, by Bohr's uniqueness theorem, the result follows.

**Lemma 2.8.6.** **Let $f \in S_L$, $L > 0$, and $f(x) < 0$ (resp. $f(x) > 0$) on a set of positive Lebesgue measure. Then there exists $\mu > 0$ (resp. $\mu < 0$) for which**

$$-y'' + \mu f(x)y = 0$$

**is not disconjugate (and so oscillatory) on R.**

Proof. See §2.9 below, remark 3 and corollary 2.7.2.

**Theorem 2.8.7.** **Let $f \in S_L$, $L > 0$, f not a.e. zero, $M\{f(x)\} = 0$. Then there exists $\mu^- < 0$, $\mu^+ > 0$ such that**

$$-y'' + \mu f(x)y = 0$$

**is oscillatory on $(-\infty, \mu^-] \cup [\mu^+, \infty)$.**

Proof. Since $f \in S_L$, $M\{f(x)\} = 0$ and f is not a.e. zero, the corollary 2.8.5 implies the existence of two sets of positive Lebesgue measure on each of which $f(x) > 0$, $f(x) < 0$. Lemma 2.8.6 implies the existence of $\mu^- < 0$, $\mu^+ > 0$ so that

$$-y'' + \mu^{\pm} f(x)y = 0$$

is oscillatory on **R**. The result now follows from the fact that $D = N$ (corollary 2.7.2) along with the convexity of $D$, (theorem 2.1.12).

**Theorem 2.8.8.** **Let $V \not\equiv 0$ be a real piecewise continuous Stepanov a.p. function with $|v(x)| \leq M$, $x \in$ R. Then (2.5.8) is oscillatory at $\pm\infty$ for every real $\lambda \neq 0$ if and only if $M\{V(x)\} = 0$.**

Proof. The sufficiency follows from case 2 of the sufficiency part of theorem 2.5.12 and the fact (theorem 2.7.1) that a non-disconjugate Stepanov a.p. equation (2.5.5) must be oscillatory at both $\pm\infty$. Since, for a Stepanov a.p. function, the mean-value (2.8.5) exists

$$\lim_{T\to\infty} \frac{1}{T} \int_a^{a+T} f(s)ds = M\{f(x)\} \tag{2.8.15}$$

(uniformly in a) and is finite [Be.1] the necessity of the proof of theorem 2.5.12 applies directly to prove the necessity of the condition above.

**Note:** An example of a Stepanov a.p. function which is not Bohr a.p. is given by

$$V(x) = \begin{cases} +1 & \text{on} \quad (0,\pi) \\ -1 & \text{on} \quad (\pi,2\pi) \end{cases}$$

and $V(x+2\pi) = V(x)$. Moreover $\sup\{|v(x)| : x \in \mathbf{R}\} < +\infty$ here.

On the other hand,

$$V(x) = \begin{cases} |x|^{-1/2}, & x \in (0,\pi) \\ |x|^{-1/2}, & x \in (\pi,2\pi) \end{cases}$$

is also Stepanov a.p. but $\sup\{|v(x)| : x \in \mathbf{R}\} = +\infty$.

**Theorem 2.8.9** **Let $V \not\equiv 0$ be a real piecewise continuous Weyl or Besicovitch a.p. function with $|v(x)| < M$, $x \in \mathbf{R}$. Then (2.5.5) is oscillatory at $+\infty$ if $M\{(v-\mu)_+^2\}M\{(v-\mu)_-^2\} > 0$ for every $\mu < \|v\|_\infty$.**

Proof. This is essentially the same as that of theorem 2.5.12(2). Note that for Weyl/Besicovitch a.p. functions there exists a mean-value (2.8.5) which is uniform in a, [Be.1]. We cannot conclude at this time whether there is oscillation also at $-\infty$ in contrast with the Stepanov a.p. case, theorem 2.8.8.

**Example:** The following construction produces an example of a Weyl a.p. function. Let $\phi(t)$ be a non-trivial Bohr a.p. function and let $\varepsilon > 0$ be given. Then there is a relatively dense set of $\varepsilon$-translation numbers $\{\tau_i\}$. Let $\delta < 1$ and define the function $K(t) = K(t,\delta)$ by setting $K(t) = 1$ on every interval of the form $\tau_i \leq t \leq \tau_i + \delta$, and $k(t) = 0$ otherwise.

Then $K(t)$ is Weyl a.p. [Be.1, p. 92; Lemma ], and its mean value, $M\{K(t)\} = \delta \lim (i/\tau_i)$ as $i \to \infty$, [Be.1, p.93; corollary 2]. Note, however, that the equation

$$y'' + K(t)y = 0, \qquad t \,\varepsilon[0, \infty),$$

is oscillatory, independently of the choice of the sequence $\tau_i$ and of the value of $\delta$. This is because

$$\int_0^\infty K(s)\, ds = \infty$$

and this, in itself implies oscillation [this is essentially due to W.B.Fite (1911) for continuous K, but this can also be extended to the present case without any difficulty.]

**Theorem 2.8.10** **Let $V \in L_\infty(\mathbf{R})$, not a.e. zero, be a Stepanov almost periodic function with $M\{V(x)\} = 0$ and having an unbounded indefinite integral, (unbounded above and below). Then**

$$y'' + V(x)y = 0, \qquad x \in \mathbf{R}$$

**is oscillatory at $\pm\ \infty$.**

Proof. This is relatively straightforward and uses the basic idea of Favard [Fa.1, p. 48]. Essentially that result i.e., lemma 2.5.9, is valid if V is merely bounded on **R** and its indefinite integral is an absolutely continuous function. Thus lemma 2.5.9 has a counterpart for bounded Stepanov a.p. functions. The proof now follows the lines of the sufficiency of theorem 2.5.12, case 1, and so is omitted.

**Example:** The function V defined by V(x) = sgn(x) on [-1,+1], where sgn(x) represents the sign of x, and extended to the whole line by means of the relation V(x+2) = V(x), is Stepanov a.p. with M{V(x)} = 0. Thus the last remark implies that

$$y'' + \lambda\, \mathrm{sgn}(x) y = 0$$

is oscillatory on $(-\infty, \infty)$ for every real $\lambda$ not zero.

## 2.9 Concluding remarks on §2.

1. We saw in §2.2 that $N$ is not generally a closed set for an equation of the form (2). It is natural to ask whether $N$ is ever closed - Of course $N$ will be closed whenever $N = D$ (see §2.7) however it is not known whether $N$ may be closed when $N \neq D$.

2. In §2.5 we presented various results relating to the equation

$$y'' + (\lambda - A(x))y = 0, \qquad (2.9.1)$$

where A is a Bohr almost-periodic function. Let us restrict (2.9.1) to $(0, \infty)$. Since A(x) is bounded below Weyl's criterion, [Gl.1] implies that (2.9.1) is **limit-point** at $+\infty$ (i.e., for each $\lambda \in C$, (2.9.1) has at least one solution $y \notin L_2(0, \infty)$). In this case the homogeneous separated boundary condition

$$y(0) \cos \alpha - y'(0) \sin \alpha = 0 \qquad (2.9.2)$$

where $\alpha \in [0, \pi)$ generates an eigenvalue problem and the operator L defined by

$$Lf = -f'' + A(x)f$$

with domain $D(L) = \{f \in L^2(0, \infty) : f, f' \in AC_{loc}(0, \infty), Lf \in L^2(0, \infty) \text{ and } f(0) \cos \alpha - f'(0) \sin \alpha = 0\}$ is self-adjoint.

Now we know that (2.9.1) has only oscillatory solutions on $[0, \infty)$ for $\lambda = M\{A\}$. Consequently for $\lambda > M\{A\}$, (2.9.1) must have only oscillatory solutions by the Sturm comparison theorem. Thus any eigenvalues appearing in the region $\lambda > M\{A\}$ would have oscillatory eigenfunctions!

3. The proof of lemma 2.1.2 may be used to prove the following result: Let q be as in the said lemma. **Then (2.1.3) is disconjugate for every $\lambda > 0$ ($\lambda < 0$) implies that $q(x) \leq 0$ ($q(x) \geq 0$) a.e. on I, (or R).** This result has the following consequences:

a) Let V: $\mathbf{R} \to \mathbf{R}$, $V \in L_1^{loc}(\mathbf{R})$. If there exists a set of positive Lebesgue measure on which $V(x) < 0$, then there exists a value of $\lambda > 0$ such that the equation

$$-y'' + \lambda V(x)y = 0, \qquad x \in \mathbf{R} \tag{2.9.3}$$

is not disconjugate on **R**.

b) This implies, **in particular, that the bottom of the spectrum of the operator**

$$\frac{-d^2}{dx^2} + \mu V(x), \qquad x \in \mathbf{R}$$

**in $L_2(\mathbf{R})$ lies in $\{\lambda \in \mathbf{R}: \lambda \in (-\infty, 0)\}$, if $\mu > 0$ is suitably chosen.** (Note that here V is not necessarily of one "sign"). The latter result may be of interest in higher dimensions cf., [Si.1; theorem B.5.3 and §C8].

## 3. Linear Vector Ordinary Differential Equations

### 3.1 Fundamental notions, definitions and terminology

**In this section A, B will denote n x n matrices over the real field R (unless otherwise specified). The symbol $A(x) \in P$ signifies that every entry of A(x), usually denoted by $a_{ij}(x)$, has the indicated property $P$. The symbol $|A|$ will stand for a, generally unspecified, norm of A. As usual $I = [0, \infty)$.**

Let $A(x) \in L_1^{loc}(I)$ and consider the differential equation

$$y'' + A(x)y = 0, \tag{3.1.1}$$

$x \in I$, for a column-vector $y \in \mathbf{R}^n$. Then the initial value problem (3.1.1), $y(0) = y_0$, $y'(0) = y_1$ where $y_i \in \mathbf{R}^n$ are given, $(i = 0, 1)$ admits a unique solution $y \in AC_{loc}(I)$ for which $y' \in AC_{loc}(I)$ and y" satisfies (3.1.1) a.e. on I. This is most easily seen by rewriting (3.1.1) as a first-order system in 2n-space and applying an existence and uniqueness theorem, [Ev.1].

**We will call (3.1.1) Wintner-disconjugate** on I provided every non-trivial solution (i.e., not the zero-vector) y(x) of (3.1.1) has the property that y(x) = 0 (the zero-vector) for at most one x-value in $[0, \infty)$.

In the event that $A(x) = A(x)^*$ (i.e., A is symmetric) the definition of "Wintner-disconjugacy" above is equivalent to the usual definition of "disconjugacy" (see e.g. [Ha.4] or [Co.1]).

Equation (3.1.1) is **strongly-disconjugate** on $[0, \infty)$ if every non-trivial solution y(x) of (3.1.1) has the property that each one of its (non-identically zero) components changes sign at most once in $(0, \infty)$.

Equation (3.1.1) is **weakly-disconjugate** on $[0, \infty)$ if (1) is Wintner-disconjugate and, in addition, every non-trivial solution y(x) such that y(0) = 0 has the property that each one of its (non-identically zero) components has no zeros in $(0, \infty)$.

**Note:** Whenever (3.1.1) is either weakly or strongly-disconjugate, it is Wintner-disconjugate. Moreover each one of these three definitions reduces to the usual one of §2 when n = 1.

**Example 1.** We show that (3.1.1) may be Wintner-disconjugate and fail to be strongly-disconjugate.

To see this let A(x) = A where A is the constant (symmetric) matrix

$$A = \begin{pmatrix} -3 & 1 \\ 1 & -3 \end{pmatrix}.$$

The the **spectrum of A,** $\sigma(A) = \{-2, -4\}$ and so (1) is Wintner-disconjugate on $[0, \infty)$, (see lemma 3.2.1 below). Writing $y(x) = \text{col}\,(y_1(x), y_2(x))$ it is readily verified that the solution of (3.1.1) given

by $$y(x) = \begin{Bmatrix} (\cosh 2 - \cosh\sqrt{2})\sinh(x\sqrt{2}) + \sinh(\sqrt{2})(\cosh x\sqrt{2} - \cosh 2x) \\ (\cosh 2 - \cosh\sqrt{2})\sinh(x\sqrt{2}) + \sinh(\sqrt{2})(\cosh x\sqrt{2} + \cosh 2x) \end{Bmatrix}$$

satisfies $y_1(0) = y_1(0) = 0$. Consequently, the resulting system is not strongly disconjugate.

**Example 2.** There are cases in which (3.1.1) may be Wintner-disconjugate and, simultaneously, weakly-disconjugate. For example let

$$A = \begin{pmatrix} 0 & 0 \\ 0 & -1 \end{pmatrix}$$

Then $\sigma(A) = \{0, -1\}$ (by lemma 3.2.1 below) and so (3.1.1) is Wintner-disconjugate. Moreover the general solution satisfying $y(0) = 0$ is given by $y(x) = \text{col}\,(\mu x, \nu \sinh x)$, $\mu, \nu \in \mathbf{R}$, neither component of which can vanish in $(0, \infty)$. Hence it is also weakly-disconjugate.

However it is also possible that (3.1.1) may be Wintner-disconjugate and not weakly-disconjugate...

**Example 3.** To see this let A be defined by

$$A = \begin{pmatrix} 0 & 0 \\ 1 & 0 \end{pmatrix}$$

Then A is nilpotent and so $\sigma(A) = \{0\}$. Consequently (3.1.1) is Wintner-disconjugate on $[0, \infty)$. The general solution of (3.1.1) satisfying $y(0) = 0$ is given by $y(x) = \cos\,(\mu x, -\mu x^3/6 + \nu x)$ where $\mu, \nu \in \mathbf{R}$. Let $\mu = 6$, $\nu = 1$. Then the corresponding solution satisfies $y(0) = 0$ but its second component $y_2(x)$ has a zero at $x = 1$. Hence (3.1.1) cannot be weakly-disconjugate.

**Note.** Whenever A is a constant symmetric matrix, the notion of strong-disconjugacy implies that of weak-disconjugacy. However, it seems as if these two notions are independent, in general.

## §3.2 Wintner-disconjugacy

Let A(t), B(t) be n x n matrix functions satisfying the usual conditions (§3.1).

Consider the vector differential equation

$$y'' + (-\alpha A(t) + \beta B(t))y = 0 \qquad (3.2.1)$$

where $\alpha, \beta \in \mathbf{R}$ and $y \in \mathbf{R}^n$ is a column vector. The problem of interest here consists in describing the large-scale properties of that region in parameter-space $\mathbf{R}^2 = \{(\alpha, \beta): \alpha, \beta \in \mathbf{R}\}$ for which (3.2.1) is Wintner-disconjugate. This region will be called the **domain of Wintner-disconjugacy** or the **Wintner domain,** for brevity: It will be denoted by $W$ ?

**Lemma 3.2.1** **Let A be a constant (real or complex) n x n matrix. Then the equation**

$$y'' + Ay = 0, \qquad x \in [0, \infty) \qquad (3.2.2)$$

**is Wintner-disconjugate on $[0, \infty)$ if and only if $\sigma(A) \cap (0, \infty) = \emptyset$.**

Proof. (Necessity). There is a non-singular matrix T such that $A = T\,U\,T^{-1}$ where $U$ is upper-triangular. The transformation $y = Tz$ reduces (2.3.2) to the system $z'' + U z = 0$ and preserves **conjugate points** [Ha.4]. Now the leading entry of $U$ may be chosen arbitrarily in the Schur form of A, and so by repeating the following procedure n times, the principal entry may be made successively equal to each of the eigenvalues of A. Note that the vector function $z(x) = \mathrm{col}\,(z_1(x), 0, ..., 0)$, where $z_1 \neq 0$ satisfies $z''_1 + a_{11} z_{11} = 0$ is, in fact, a solution of $z'' + Uz = 0$. Since the latter is Wintner-disconjugate either $a_{11}$ is strictly complex or $a_{11} \le 0$. The above considerations similarly show that either $a_{ii}$ is strictly complex or $a_{ii} \le 0$. Hence $\sigma(A) \cap (0, \infty) = \emptyset$.

Conversely assume, on the contrary, that (3.2.2) is not Wintner-disconjugate on $[0, \infty)$ and that $\sigma(A) \cap (0,\infty) = \emptyset$. Then there exists two points $x_1$, $x_2$ and a solution $y(x) \neq 0$ of (3.2.2) such that $y_1(x) = y(x_2) = 0$ (the zero vector). We now proceed as in the necessity, reduce A to its Schur form and consider $z'' + Uz = 0$. Since T is non-singular $z(x_1) = z(x_2) = 0$ for some $z \neq 0$ satisfying the latter. For $z(x) = \mathrm{col}\,(z_1(x), ..., z_n(x))$ note that $z_n(x)$ satisfies $z''_n + a_{nn}z_n = 0$, $z_n(x_1) = z_n(x_2) = 0$. Thus $a_{nn}$ must be an eigenvalue of the preceding boundary problem. However this is impossible since $a_{nn} \notin (0,\infty)$. Thus $z_n(x) \equiv 0$. We now proceed, "up the diagonal". Note that $z_{n-1}(x)$ satisfies

$$z''_{n-1} + a_{n-1,\, n-1}\, z_{n-1} + a_{n-1,\, n}\, z_n = 0.$$

Since $z_n \equiv 0$, we find that

$$z''_i + a_{ii}\, z_i = 0, \qquad x \in I$$

for $i = n-1$. Once again $z_i(x_1) = z_i(x_2) = 0$, for such an i. Thus repeating the argument above we find $z_{n-1} = 0$. Continuing in this way we eventually arrive at $z_i(x) \equiv 0$, $i = 1, 2, ..., n$, i.e., $z(x) \equiv 0$, $x \in I$ which is a contradiction.

### §3.2.A The case of general matrix coefficients

In contrast with the results in §2, the Wintner domain $W$ is not closed in general. The following example illustrates this.

**Example 5.** Define A, B as follows.

$$A = \begin{pmatrix} 0 & -1 \\ 0 & 0 \end{pmatrix} \qquad\qquad B = \begin{pmatrix} 6 & 16 \\ -1 & -2 \end{pmatrix}$$

Consider the equation (3.2.1) for $x \in I$. The point $(\alpha, 1) \in W$ when $\alpha > 0$, since the eigenvalues of $-\alpha A + B$ for $\alpha > 0$ are purely complex, (lemma 3.2.1). Now let $(\alpha_n, 1) \in W$ be a sequence such that $\alpha_n > 0$ and let $(\alpha_n, 1) \to (0, 1)$ in $\mathbf{R}^2$. Then $(0, 1) \notin W$ since $\sigma(B) = \{2\}$, (lemma 3.2.1). Thus $W$ cannot be a closed set (in the usual topology).

Similarly the next example shows that $W$ **is not convex in general.**

**Example 6.** Define A, B as follows.

$$A = \begin{pmatrix} 1 & -1 \\ 1 & 1 \end{pmatrix} \qquad\qquad B = \begin{pmatrix} 1 & 1 \\ -1 & 1 \end{pmatrix}$$

Note that $B = A^{-1}$. In this case both $(-1, 0), (0, 1) \in W$ (by lemma 3.2.1). If $W$ is convex then

$$y'' + (\gamma A + (1-\gamma)B)y = 0$$

is Wintner-disconjugate for each $\gamma \in [0, 1]$. However for $\gamma = 1/2$, $\sigma(\gamma A + (1-\gamma)B) = \{1\}$ and so $(-1/2, 1/2) \notin W$. On the other hand this point lies on the line segment joining $(-1, 0)$ to $(0,1)$. Thus $W$ cannot be convex.

**Remark 1. It cannot be concluded that $W$ is never convex in the case of non-symmetric coefficients.** A particular result in this direction is the following.

**Lemma 3.2.2. Let A, B be constant, commuting real n x n matrices each having purely imaginary spectrum (i.e., $\sigma(A)$, $\sigma(B) \subset \{z \in C: \ \mathrm{Re}\ z = 0\}$). Then $W$ is convex.**

Proof. Since A, B commute they are simultaneously upper-triangulable, i.e., there exists a non-singular matrix T such that $A = TUT^{-1}$. $B = TVT^{-1}$, and each one of $U$, $V$ is upper-triangular, [Ho.1]. The transformation $y = Tz$ preserves conjugate points and reduces (3.2.1) to

$$z'' + (-\alpha U + \beta V)\, z = 0.$$

Note that $W$ is invariant under the above transformation. Let $(\alpha,\beta)$, $(\alpha^*, \beta^*) \in W$. The condition for convexity is that the equation

$$z'' + (-(\gamma\alpha+(1-\gamma)\alpha^*)U + (\gamma\beta+(1-\gamma)\beta^*)V)z = 0 \qquad (3.2.3)$$

be Wintner-disconjugate for each $\gamma \in [0, 1]$. Note that the quantities $\gamma, \alpha, \beta, \alpha^*, \beta^* \in \mathbf{R}$ so that the spectrum of the matrix coefficient appearing in (3.2.3) is once again purely imaginary (may contain zero). Hence lemma 3.2.1 applies showing that (3.2.3) is Wintner-disconjugate for each $\gamma \in [0, 1]$. The result follows from this.

**Remark 2.** A direct analog of lemma 2.1.2 is not available in this general setting. For example, let R be the nilpotent matrix

$$R = \begin{pmatrix} 0 & 1 \\ 0 & 0 \end{pmatrix}.$$

Then $y'' + \lambda Rr = 0$ is, in fact, Wintner-disconjugate for **each real $\lambda \neq 0$.** However $R \neq 0$ (the zero matrix).

Because of the failure of an analog of the said lemma many of the results in §2 have no clear counterpart in the case of (3.2.1) with general matrix coefficients. However, the situation of symmetric matrix coefficients is markedly similar to that in §2 as we shall now see.

### §3.2.B The case of symmetric matrix coefficients

We will always assume that $A = A^*$, $B = B^*$ in (3.2.1). We retain the notation of the previous section and, as usual, $I = [0, \infty)$.

**Lemma 3.2.3.** **Let $P, Q \in L_1^{loc}(I)$ with $P(x) = P(x)^*$, $Q(x) = Q(x)^*$ for $x \in I$. If each of the equations $y'' + P(x)y = 0$, $y'' + Q(x)y = 0$, is Wintner-disconjugate on I, then**

$$y'' + (\gamma P(x) + (1-\gamma)Q(x)y = 0 \qquad (3.2.4)$$

**is also Wintner-disconjugate on I for each $\gamma \in [0, 1]$.**

Proof. Let $\mathcal{A}_1(a, b) = \{\eta: [a, b] \to \mathbf{R}^n \mid \eta \in AC[a, b], \eta' \in L_2(a,b) \text{ and } \eta(a) = \eta(b) = 0\}$. Then it is known [Ha.4] or [Re.1] that $y'' + Q(x)y = 0$ is Wintner-disconjugate on I if and only if for every closed and bounded interval $[a, b] \subset I$ the functional

$$I(\eta,P; a,b) \equiv \int_a^b [\eta'(t)\bullet\eta'(t) - P(t)\eta(t)\bullet\eta(t)]dt$$

is positive definite on $\mathcal{A}_1(a, b)$. The proof now follows the presentation in lemma 2.1.9 and so is omitted.

**Theorem 3.2.4** **Let $A, B \in L_1^{loc}(I)$. Then the Wintner domain of (3.2.1) is convex.**

Proof. This is similar to the proof of theorem 2.1.12 except that now lemma 3.2.3 is used. Since the arguments can be easily reconstructed we shall omit them.

Of peripheral interest is the following result.

**Lemma 3.2.5.** **Let A, B be constant real symmetric matrices and assume that**

$$\inf \{|x^* (A + iB)x| : x \in \mathbf{R}^n, \ \|x\| = 1\} > 0$$

**for n ≥ 3. Then *W* contains at least one finite line segment through (0, 0).**

Proof. The hypothesis imply the existence of a number $\theta \in \mathbf{R}$ for which $x^* (A \sin \theta + B \cos \theta)x > 0$ for each $x \in \mathbf{R}^n$, $n > 3$, and $x \neq 0$, [Sr.1, theorem 2.2]. That is, the matrix $A \sin \theta + B \cos \theta$ is positive definite for such θ. Thus let $\alpha = \sin \theta$, $\beta = -\cos \theta$. Then the matrix $\alpha A - \beta B$ is positive definite. But the latter implies that (3.2.1) is Wintner-disconjugate on I (a conseuqence of Morse's extension of the Sturm comparison theorem; see e.g. [Re.1]). Hence $(\sin \theta, -\cos \theta) \in W$. The result now follows upon an application of theorem 3.2.4.

**Lemma 3.2.6. Let $A_m = A^*_m$ be a sequence of continuous matrix functions on I such that for some symmetric matrix A, we have for each compact subset [a, b] in I**

$$\lim_{x \to \infty} \int_a^b |A_m(x) - A(x)| dx = 0.$$

**If for each m,**

$$y'' + A_m(x)y = 0 \tag{3.2.5}$$

**is Wintner-disconjugate on I, then**

$$y'' + A(x)y = 0 \tag{3.2.6}$$

**is also Wintner-disconjugate on I.**

Proof. (3.2.5) is Wintner-disconjugate on I if and only if for every $\eta \neq 0$ in $A_1(a,b)$, and every compact subset [a, b] in I we have

$$\int_a^b (\eta'(t)\bullet\eta'(t) - A_m(t)\eta(t)\bullet\eta(t))dt \geq 0$$

with equality holding if and only if $\eta = 0$. Now given [a, b] and $\eta \neq 0$ in $A_1(a,b)$, we have for a given $\varepsilon > 0$.

$$\int_a^b (\eta'(t)\bullet\eta'(t) - A(t)\eta(t)\bullet\eta(t))dt \geq \int_a^b \eta(t)\bullet(A_m(t) - A(t))\eta(t)dt$$

$$\geq -\sup_{t\in[a,b]} |\eta(t)\cdot\eta(t)| \cdot \int_a^b |A_m(t)-A(t)|dt$$

$$\geq -\varepsilon$$

provided m is sufficiently large. We may now let $m\to\infty$ to find

$$I(\eta,A;a,b) \geq 0$$

Hence this is true for each $\eta \in \mathcal{A}_1(a, b)$ and consequently for each $\eta \in \mathcal{A}_1(a, b)$ and every [a, b] in I. But this now implies vis Jacobi's theorem [Ha.4, exercise 10.1] that y" + A(x)y = 0 is disconjugate on I.

**Remark 3.** **It is very likely that the continuity hypothesis on $A_m$ and A may be weakened to merely $L_1^{loc}(I)$. In this respect see [Re.2, §V.6]**

**Theorem 3.2.7.** **Let A, B be continuous on I. Then the Wintner domain of (3.2.1) is a closed set.**

Proof. This uses lemma 3.2.6 and an argument analogous to that presented in theorem 2.1.11.

We may now reformulate most of the results in §2.1 by using the theory of positive linear functionals on the vector space $S$ **of all real symmetric n x n matrices** as considered by various authors [Et.1], [Ha.5].

A **linear functional** g: $S \to \mathbf{R}$ is a map such that g(A + B) = g(A) + g(B), and g(cA) = cg(A) for each A, B $\in S$, c $\in \mathbf{R}$. A linear functional g is **positive** if g(A) ≥ 0 whenever A ≥ 0 (i.e. A is positive semidefinite). We refer the reader to [Et.1] and [Ha.5] for other related properties of such functionals. In particular the representation theorem for g implies that g(a(x)) will be Lebesgue measurable for Lebesgue measurable A $\in S$. Moreover, if A $\in L_1^{loc}$(I) the same is true of g(A).

**Lemma 3.2.8.** **Let A $\in S \cap L_1^{loc}(I)$. If (3.2.6) is Wintner-disconjugate on I and g is an arbitrary (non-trivial) positive linear functional on $S$ then the scalar equation**

$$y'' + g(A(x))y = 0 \qquad (3.2.7)$$

**is disconjugate on I.**

Proof. Fix $[a, b] \subset I$. Since (3.2.6) is Wintner-disconjugate on I it is *a fortiori* Wintner-disconjugate on [a, b]. Hence [Ha.5], (3.2.7) is disconjugate on [a, b]. But [a, b] is arbitrary. Hence, for fixed g, (3.2.7) is disconjugate on I. Since g is arbitrary the conclusion follows.

**Lemma 3.2.9.** **Let $A \in S$ have the property that for every positive linear functional $g: S \to \mathbf{R}$, we have $g(A) = 0$. Then $A = 0$, the zero matrix in $S$.**

Proof. Let $e_i$ be the usual unit vectors in $\mathbf{R}^n$ having a "1" in the $i^{th}$ place and zeros elsewhere. Let g denote the positive linear functional defined by

$$g(A) \equiv e_i^* A e_i = a_{ii}$$

where $a_{ii}$ is the $i^{th}$ diagonal entry of A. By hypothesis $g(A) = 0$. Hence $a_{ii} = 0$, for $i = 1, 2, \ldots, n$. Generally let $e_{ij}$ denote the vector in $\mathbf{R}^n$ which has a "1" in its $i^{th}$ and $j^{th}$ coordinates, where $i \leq j$, and 0 elsewhere. Then

$$g(A) \equiv e_{ij}^* A e_{ij}$$

is again a positive linear functional on $S$. But $e_{ij}^* A e_{ij} = 2a_{ij}$ since A is symmetric. Hence $a_{ij} = 0$ for $1 \leq i, j \leq n$ and the result follows.

**Remark 4.** The result in [Ha.5] alluded to during the course of the proof of lemma 3.2.8 is stated for continuous A(x) - However, the techniques in [Re.1, Re.2] allow the extension of the former result to this more general setting.

We are now in position to prove a analog of lemma 2.1.2 in the vector case.

**Theorem 3.2.10** **Let $Q \in S \cap L_1^{loc}(I)$. If the equation**

$$y'' + \lambda Q(x)y = 0, \qquad x \in I$$

**is Wintner-disconjugate for each $\lambda$, $-\infty < \lambda < +\infty$, then $Q(x) = 0$ (the zero matrix) a.e. on I.**

Proof. Let g be a arbitrary but fixed positive linear functional on $S$. Then lemma 3.2.8 implies that the scalar equation

$$y'' + \lambda g(Q(x))y = 0, \qquad x \in I \tag{3.2.8}$$

is disconjugate on I, for each $\lambda \in \mathbf{R}$. The same lemma implies, in fact, that the latter holds for **every** positive linear functional g. Hence lemma 2.1.2 applies yielding $g(Q(x)) = 0$, a.e for $x \in I$. But this holds for every g. Hence lemma 3.2.9 implies that $Q(x) = 0$ a.e., on I.

The above theorem 3.2.10 now makes possible an extension of most of results in §2 from the scalar to the vector case. We will only mention here some of the principal results which ensure from the above results and leave the proofs to the reader (as they are, in all respects, similar to those in §2). The first result is a reformulation of theorem 3.2.10.

**Theorem 3.2.11** **a) Let $Q \in \mathcal{S} \cap L_1^{loc}(I)$. If $Q(x) \neq 0$ (the zero matrix) on a set of positive Lebesgue measure in I, then there exists a value of $\lambda \in R$ such that**

$$y'' + \lambda Q(x)y = 0 \tag{3.2.9}$$

**is not Wintner-disconjugate on I.**

**b) Let $Q(x) > 0$ (be positive definite, in the form sense) on a set of positive Lebesgue measure in I. Then there exists a value of $\lambda > 0$ that (3.2.9) is not disconjugate on I.**

**Remark 5.** Theorem 3.2.11(b) is in the same spirit as remark 3 in §2.9. It is closely related to a result of Hawking and Penrose [Hw.1, p. 541] in cosmology. We note once more that I, in the above theorem, may be replaced by **R**.

**Theorem 3.2.13.** ***W*** **contains a proper subspace of the vector space $R^2$ (other than the subspaces $\alpha = 0$ and $\beta = 0$), if and only if the matrices A, B are linearly dependent.**

The former result is the analog of theorem 2.1.5 while the latter extends theorem 2.1.7. Let us now give a analog of theorem 2.3.1.

**Theorem 3.2.14. Let A, B $\in S \cap L_1^{loc}(I)$ and let g be some positive linear functional on $S$ such that**

$$\lim_{x\to\infty} g\{\int_0^x A(s)ds\} = +\infty$$

**and**

$$-\infty < \liminf_{x\to\infty} g\{\int_0^t\int_0^x B(s)ds\,dx\} \leq \limsup_{x\to\infty} g\{\int_0^t\int_0^x B(s)ds\,dx\} < \infty.$$

**Then $W \subset \{(\alpha,\beta) \in R^2 : \alpha > 0\} = H^+$.**

Proof. This is a direct consequence of theorem 4.3 in [Et.1] as applied to a analogous proof of theorem 2.3.1.

**Theorem 3.2.15. Let P(x) = P(x+1) be a continuous symmetric, real-valued matrix function of mean value equal to zero, i.e.,**

$$\int_0^1 P(s)ds = 0$$

**(the zero matrix). Then the Wintner-domain of the equation**

$$y'' + (-\alpha I + \beta P(x))y = 0$$

**is contained in $\{(\alpha, \beta): \alpha > 0\} \cup \{(0, 0)\}$**

Proof. Let g be a nontrivial positive and normalized (i.e., g(I) = 1) linear functional on $S$ and define G(x) = g(P(x)). Then G is continuous, periodic of period 1, and M{G} = 0 since g is linear. Moreover the scalar equation

$$y'' + (-\alpha + \beta G(x))y = 0$$

is oscillatory whenever $\alpha \leq 0$ and $\beta \neq 0$, [Mo.1]. The same is therefore true of the vector equation [Et.1, theorem 4.3] or [Ha.5]. The result follows.

**Note** In the proofs of the last two theorems, (3.2.1) is (Wintner) **oscillatory** if for each X > 0 there exists a nontrivial solution y(x) of (3.2.1) such that (3.2.1) is not Wintner-disconjugate on [X,∞)

## §3.3 Strong and weak disconjugacy

For basic notation and terminology refer to §3.1. The **domain of strong-disconjugacy**, *P*, consists of the collection of all pairs $(\alpha, \beta) \in \mathbf{R}^2$ such that (3.2.1) is strongly- disconjugate. In the same way we define the domain of **weak-disconjugacy**, *F*.

**Theorem 3.3.1** **Let A, B $\in L_1^{loc}(I)$ be continuous (and not necessarily symmetric) matrices. Then *P* (resp. *F*) is non-empty and *P* (resp. *F*) is a closed set.**

Proof. Note that $(0, 0) \in P \cap F$. Now let $(\alpha_i, \beta_i)$ be a sequence of points in *P* such that $(\alpha_i, \beta_i) \to (\alpha, \beta)$ in $\mathbf{R}^2$. Then

$$(-\alpha_i A(x) + \beta_i B(x)) \to (-\alpha A(x) + \beta B(x))$$

in L(a, b) where $[a, b] \subset I$.

Assume, if possible, that $(\alpha, \beta) \notin P$. Then there is a solution y(x) of (3.2.1), $y \not\equiv 0$, with the property that at least one of its, non-identically zero, components changes sign at least twice in I. Hence y has at least two zeros $x_1$, $x_2$ in $[0,\infty)$ about each of which the component changes its sign.

Now let $y_i(x)$ be the solution of

$$y'' + (\alpha_i A(x) + \beta_i B(x))y = 0$$

be such that $y_i(x_1) = 0$, $y_i'(x_1) = y'(x_1)$. Then lemma 3.2.6 implies that, for given $\varepsilon > 0$, and some fixed $\delta > 0$,

$$\sup_{[x_1,x_2+\delta]} |y_i(x) - y(x)| < \varepsilon$$

for all sufficiently large i. Let $y^{(j)}(x)$ be the exceptional component of y for which $y^{(j)}(x_1) = y^{(j)}(x_2) = 0$. Then

$$\sup_{[x_1,x_2+\delta]} |y_i^{(j)}(x) - y^{(j)}(x)| < \varepsilon \tag{3.3.1}$$

where $y_i^{(j)}$ is the $j^{th}$ component of $y_i$. Since $y^{(j)}(x)$ changes sign about $x = x_2$, (3.3.1) shows that $y^{(i)}(x)$ must equal zero near $x = x_2$ if i is sufficiently large. A similar argument applies near $x = x_1$, so that $y_{(j)}(x)$ must change sign there also. Hence $y^{(j)}(x)$ changes sign at least twice and so $(\alpha, \beta) \notin P$. A analogous proof applies, with minor changes, for *F*.

**Lemma 3.3.2.** **a) Let A be a weighted permutation matrix (i.e., a matrix for which each row and each column has precisely one non-zero entry). Then A = PD where A is a permutation matrix and D is diagonal.**

**b) Let A be diagonalizable, $A = TDT^{-1}$ where D is diagonal. If T is a weighted permutation matrix then A must be diagonal.**

Proof. a) Let $A = (a_{ij})$ and $P = (p_{ij})$ be defined by

$$P_{ij} = \begin{cases} 1 & \text{if } a_{ij} \neq 0 \\ 0 & \text{otherwise.} \end{cases}$$

The P is a permutation matrix and $D = (d_{ij})$ is given by $d_{jj} = a_{ij}$, whenever $a_{ij} \neq 0$. The result follows.

b) Let $A = TDT^{-1}$. Then by (a), $T = PD^*$ where P is a permutation matrix and $D^*$ is diagonal. Since T is invertible, so is $D^*$ and thus $T^{-1} = D^{*-1}P^{-1}$. Hence $A = (PD^*)\,D\,(D^{*-1}P^{-1}) = P(D^*D\,D^{*-1})P^{-1}$. Now $D^*D\,D^{*-1}$ is necessarily diagonal and P (and so $P^{-1}$) is a permutation matrix. Thus A must be diagonal as one can readily see.

We say that the equation $y'' + Ay = 0$ is **very strongly-disconjugate** if every non-trivial solution y has the property that every one of its (non-identically zero) components has at most one zero in $[0,\infty)$.

It is clear from the definitions that a very strongly-disconjugate equation is strongly-disconjugate. In fact,

**Theorem 3.3.3.** **Let A be a diagonalizable constant matrix with simple spectrum. Then**

$$y'' + Ay = 0$$

**is very strongly-disconjugate on I if and only if $\sigma(A) \cap (0,\infty) = \emptyset$ and A is, itself, diagonal.**

Proof. Let A be diagonalizable, $A = TDT^{-1}$ where T is nonsingular and D is diagonal. Assume that (3.2.2) is very strongly-disconjugate. Let $y = Tz$ transform (3.2.2) into a diagonal system

$$z'' + Dz = 0, \qquad x \in I. \tag{3.2.2}$$

Then $z(x) = \text{col}(z_1(x), ..., z_n(x))$ is a solution of (3.3.2) if and only if $z''_i + \lambda_i(A)z_i = 0$, $i = 1, 2, ..., n$, where $\lambda_i(A)$ represents the eigenvalues of A. Now (3.2.2) is very strongly-disconjugate, hence, it is Wintner disconjugate and so, lemma 3.2.1 implies that $\sigma(A) \cap (0, \infty) = \emptyset$. Now $y(x) = \text{col}(y_1(x), ..., y_n(x))$ will satisfy

$$y_i(x) = \sum_{j=1}^{n} t_{ij} z_i(x), \quad i = 1, 2, ..., n.$$

where $T = (t_{ij})$ is defined above. Fix i, $1 \le i \le n$, and assume that $t_{ik}, t_{il} \neq 0$, for $1 \le k \le l \le n$. Then

$$y_j(x) = t_{ik}\, z_k(x) + t_{il}\, z_l(x) + (\text{other terms}) \tag{3.3.3}$$

But each of $z_k$, $z_l$ is a linear combination of two linearly independent basis elements for the solution space of $z''_j + \lambda_j(A)z_j = 0$, $j = k, l$. Let us set the other terms in (3.3.3) equal to zero (i.e., $z_j(x) \equiv 0$ for $j \neq k$, $j \neq l$, $1 \le j \le n$). There are then **four** arbitrary constants appearing in (3.3.3). Hence one can choose these constants so that $y_i$ will have a zero at two prescribed points. Thus (3.2.2) cannot be very strongly-disconjugate which is a contradiction. Hence $t_{ik} \neq 0$, $t_{il} \neq 0$ for $1 \le k < l \le n$ is impossible and thus, since i was arbitrary, every row of T must have at most one non-zero entry and, since T is non-singular, the same must be true of every column. It now follows that T must be a weighted permutation matrix. However, since D is diagonal, lemma 3.3.2(b) implies that A is diagonal. This proves the necessity. The sufficiency is clear and so will be omitted.

**Remark 1.** **It is not known at this time** whether or not the domain of very strong disconjugacy is closed.

2. However, note that the domain of very strong-disconjugacy of (3.2.1) **may be convex** in some cases: For example, if A, B are constant matrices each one of which is a non-positive multiple of the identity matrix.

3. Other notions of disconjugacy for general (i.e., non-symmetric) systems weaker than those presented here, will be discussed in a later work.

For example, (3.2.2) will be called **Type 1 - disconjugate** if there exists a non-trivial solution whose first component (assumed non-identically zero) is disconjugate on $[0,\infty)$. Whenever A is a constant (not necessarily symmetric) matrix such that $\sigma(A) \cap \{C\backslash(0,\infty)\} = \emptyset$ then it can be shown, using the techniques of this section, that the said equation is, in fact, type 1 - disconjugate on $[0,\infty)$.

4. The use of positive linear functionals to deal with the problems of this section was restricted to the second order (finite-dimensional) vector case for the sake of clarity.

However, the main results in [Et.1] which were used here are valid for the more general calss of differential operators in which the corfficients are themselves self-adjoint operators in the B*-algebra of bounded linear operators from a Hilbert space into itself. It appears as if the relevant results in this section (in particular, the symmetric case), may be extended to this more general setting without too much effort. We choose not to delve into this matter here.

## §3.4 Almost Periodic vector systems

The main result of this section is the formulation of a systems analog of Theorem 2.5.12 dealing with the characterization of all those (Bohr) almost-periodic symmetric matrices (i.e., all of whose entries are almost-periodic) for which the system of second order equations, $y \in \mathbf{R}^n$, $(n>1)$,

$$-y'' + \lambda V(x)y = 0, \qquad x \in I = [a,\infty), \tag{3.4.1}$$

is oscillatory at $+\infty$ for **every real** $\lambda \neq 0$. The method used parallels the presentation in §2.5 although one must now formulate a systems analog of Wintner's theorem, [Wi.1]. Indeed, we will prove a systems analog of a classical theorem by Hille [He.1, p. 243], and its (scalar) extension by Taam [Ta.1, p.495] thereby producing a method for constructing non-oscillatory vector systems. To this end, we adapt a fixed point argument used in [Mi.2, theorem 2.1.2., p. 36] to this setting and finally obtain Wintner's theorem as a consequence.

We say that the equation

$$y'' + V(x)y = 0, \qquad x \in I, \tag{3.4.2}$$

is **oscillatory** at $+\infty$ provided for each $x_0 \in I$, there is $x_1 > x_0$ such that (3.4.2) is not (Wintner) disconjugate on $[x_0, x_1]$. It is said to be **non-oscillatory** otherwise. Thus, the oscillation of (3.4.2) is equivalent to the existence, in I, of arbitrarily large conjugate points (not necessarily all belonging to one solution). Associated with (3.4.2) is the matrix equation,

$$Y'' + V(x)Y = 0, \qquad x \in I, \tag{3.4.3}$$

where Y(x) is an nxn matrix. A solution of (3.4.3) is said to be **non-trivial** if $\det Y(x) \neq 0$ for at least one $x \in I$ and a non-trivial solution Y(x) is said to be **prepared or self-conjugate** if

$$Y^*(x)Y'(x) - Y^{*\prime}(x)Y(x) \equiv 0, \qquad x \in I.$$

(It is readily shown that the left-side of the last equation is always independent of x for a given solution Y). Now (3.4.3) is said to be **oscillatory** at $+\infty$ in case the determinant of every non-trivial prepared solution vanishes on $[b, \infty)$ for every $b > 0$. It is known that (3.4.3) is oscillatory if and only if (3.4.2) is oscillatory, (see e.g. [BEM.1, p. 263]). In the sequel we assume, for simplicity, that V(x) is continuous on I (although this may be weakened to $V \in L_1^{loc}(I)$, if necessary).

**Lemma 3.4.1** **(see Re.2, theorem 6.3, p. 284]) Let a matrix Riccati system associated with (3.4.2), viz.,**

$$W'(x) + W^2(x) + V(x) = 0 \tag{3.4.4}$$

**have an absolutely continuous and symmetric (matrix) solution *W* on some closed half-line contained in I. Then (3.4.2) is non-oscillatory at $+\infty$, (actually, disconjugate on said half-line).**

Proof. By hypothesis, W exists and is symmetric on every $[x_0, x_1]$ where $x_0$ is sufficiently large and $x_1 > x_0$. Thus [Re.2, theorem 6.3] implies that (3.4.2) is disconjugate on $[x_0, x_1]$ for $x_0$ sufficiently large and every $x_1 > x_0$. The result now follows by the definition.

**Corollary 3.4.2.** **Assume that V(x) has the property that**

$$\lim_{T\to\infty} \int_x^T V(s)ds \tag{3.4.5}$$

**exists, for each sufficiently large x, and that the limit-matrix has finite norm, for said x.**

**Let the Riccati matrix integral system associated with (3.4.2), viz.,**

$$W(x) = \int_x^\infty V(s)ds + \int_x^\infty W^2(s)ds \tag{3.4.6}$$

**have an absolutely continous and symmetric matrix solution on some closed half-line $[x_0, \infty)$, whose norm, $\|W\|$, (fixed but unspecified), is in $L^2(x_0, \infty)$. Then (3.4.2) is non-oscillatory at $+\infty$.**

Proof. This is clear form lemma 3.4.1 since, if W satisfies (3.4.6) for $x > x_0$ then W satisfies (3.4.4) on $(x_0, \infty)$ and the result follows.

**Note:** a) It is not difficult to show that, conversely, if (3.4.5) holds and (3.4.2) is non-oscillatory at $+\infty$, then (3.4.6) has an absolutely continuous and symmetric solution with

$$\left\| \int_x^\infty W^2(s)ds \right\| < \infty$$

for all sufficiently large x. We leave this fact as an exercise to the reader.

b) Furthermore, we note that, in corollary 3.4.2, we may state the conclusion more precisely as "disconjugate on $[x_0, \infty)$" rather than "non-oscillatory at $+\infty$".

The next result deals with a sufficiency criterion (of Arzelà-Ascoli type) for a collection of vector functions to be compact in $L^2(\mathbf{R})$, in the spirit of an earlier result by M. Riesz [Rz.1, p. 137], which we state here without proof for reference purposes.

**Lemma 3.4.3:** [Rz.1] **A family $F$ of functions in $L^p(\mathbf{R})$, $p \geq 1$, is compact if and only if**

**a) There is an $M > 0$ such that, for every $f \in F$,**

$$\|f\|_p \leq M$$

**b) For any $\varepsilon > 0$, there is a $\delta(\varepsilon) > 0$ such that for every $f \in F$,**

$$\|f(x+h)-f(x)\|_p < \varepsilon$$

**whenever $|h| < \delta$,**

**c) If $E_A = \{x \in \mathbf{R}: |x - x_0| > A, x_0 \text{ fixed}\}$, then, for every $f \in F$,**

$$\lim_{A\to\infty} \|f\|_{E_A} = 0$$

where the norm is the induced $L^p$-norm on $E_A$.

**Lemma 3.4.4** For $v: R \to R^{2n}$, $n \geq 1$, we define $\|v\|$, for definiteness by

$$\|v(x)\| = \sup_i |v_i(x)|$$

where $v(x) = col(v_1(x),\ldots, v_{2n}(x))$.

Let $F$ be a family of vector functions in $L^2_{2n}(R) \equiv \{v: R \to R^{2n} |\ \| \|v\| \|_2 < \infty\}$ satisfying

a) There is an $M > 0$ such that for every $v \in F$,

$$\int_R \|v(x)\|^2 dx \leq M$$

b) For any $\varepsilon > 0$, there is a $\delta(\varepsilon) > 0$ such that, for every $v \in F$,

$$\int_R \|v(x+h)-v(x)\|^2 dx < \varepsilon$$

c) If $E_A = \{x \in R: |x-x_0| > A,\ x_0 \text{ fixed}\}$, then for every $v \in F$,

$$\lim_{A\to\infty} \int_{E_A} \|v(x)\|^2 dx = 0.$$

Then $F$ is compact in $L^2_{2n}(R)$.

Proof. (Idea). The conditions (a), (b), (c) above imply that hypotheses (a), (b), (c) of lemma 3.4.3 are satisfied for f replaced by $v_i(x)$, $(1 \leq i \leq 2n)$, a fixed, but otherwise arbitrary component of v. It follows that the family $F_i$ of $i^{th}$ components of every member of $F$ is compact. Thus, for any bounded sequence $\{v^m\}$ in $F$, the sequence of $i^{th}$ components $\{v_i^m(x)\}$ has a convergent subsequence for each $i$, $1 \le i \le 2n$. Starting with $i = 1$, we specify a subsequence, which we rewrite as $v_1^m$ again, which converges. For m varying over these (latter) indices, the sequence $\{v_2^m(x)\}$ also has a convergent subsequence. We continue in this way until we reach $v_{2n}^m(x)$. The result now follows.

**Note:** Although we have stated lemma 3.4.4 in the case $p = 2$, for reasons to be made clear later, the proof is also valid for arbitrary $p \ge 1$, with the necessary changes in hypotheses (a), (b), (c) therein.

**Corollary 3.4.5: (see [Mi.2; corollary II.1.2, p. 281]) Let $F$ be a family of vector-functions defined on the closed half-line, $[a, \infty)$, each member of which is in $L^2_{2n}[a, \infty)$ and satisfying the following conditions:**

**a) There is an $M > 0$ such that, for every $v \in F$,**

$$\int_a^\infty \| v(x) \|^2 dx \le M$$

**b) If $E_A = \{x: A \le x < \infty\}$, then, for given $\varepsilon > 0$, and for every $v \in F$,**

$$\int_{E_A} \| v(x) \|^2 dx < \varepsilon$$

**for A sufficiently large.**

**c) For $\varepsilon > 0$, there is a $\delta(\varepsilon) > 0$, such that, for every $v \in F$,**

$$\int_a^\infty \| v(x+h)-v(x) \|^2 dx < \varepsilon$$

**whenever $|h| < \delta$. Then $F$ is compact in $L^2_{2n}[a,\infty)$.**

Proof: This follows essentially from lemma 3.4.4 by "extending" our class $F$ so that every member of $F$ is a.e. equal to zero on $(-\infty, a)$.

**Lemma 3.4.6 Let $F$ be a family of symmetric nxn matrices defined on $[a, \infty)$ and such that, for every member $V \in F$,**

$$\| V(x) \| \equiv \sup_{i,j} |V_{ij}(x)|$$

**and $\| V \| \in L^2[a, \infty)$. In addition, we assume that conditions (a), (b), (c) of lemma 3.4.5 are satisfied for every $V \equiv v \in F$. The $F$ is compact.**

Proof. This follows essentially by corollary 3.4.5 .

**Note:** The choice of norm used in lemmata 3.4.5 and 3.4.6 is of no particular importance and any (vector and compatible matrix-) norm can be used there.

**Theorem 3.4.7: Let $h: S \to R^+$ satisfy $h(0) = 0$ and assume that h is continuous at "0" (relative to the norm in lemma 3.4.6). Let $Q_i \in S$, $i = 1, 2$, be such that**

$$\lim_{T\to\infty} \int_x^T Q_i(s)ds \equiv \int_x^\infty Q_i(s)ds$$

**exists and is a matrix of finite norm on $S$.**

**If, for every sufficiently large x, we have**

$$h\left(\int_x^\infty Q_1(s)ds\right) \geq \| \int_x^\infty Q_2(s)ds \|$$

**and the scalar Riccati integral equation**

$$v_1(x) = h\left(\int_x^\infty Q_1(s)ds\right) + \int_x^\infty v_1^2(s)ds$$

**has a real-valued solution defined for all sufficiently large x and for which $v_1$ is in $L^2$ at $\infty$, then the matrix Riccati integral equation**

$$v(x) = \pm \int_x^\infty Q_2(s)ds + \int_x^\infty v^2(s)ds$$

**has a solution $v(x) = v^T(x)$ satisfying**

$$\int_x^\infty \| v(s) \|^2 ds < \infty$$

**if x is sufficiently large.**

Proof: We note that the continuity assumption on the functional h at $0 \in \mathcal{S}$ along with the hypothesis on the integral of $Q_1$ together imply that

$$h\left(\int_x^\infty Q_1(s)ds\right) < \infty$$

for all sufficiently large x. The method of proof now parallels a former Schauder argument [Mi.2; theorem 2.1.2, p. 37]. We consider the Banach space $L^2(I) = \{v: I \to \mathcal{S} \mid \| \, \|v\| \, \|_2 < \infty\}$ where $I \equiv [a, \infty)$, of real matrix-valued functions defined on I and whose matrix-norm is square-integrable at $\infty$. Define a subset X by

$$X = \left\{ v \in L^2(I): \ v(x) = v^T(x), \ \|v(x)\| \le v_1(x), \text{ all large } x \right\}$$

For $v \in X$ we define a map T by

$$(Tv)(x) = \int_x^\infty Q_2(s)ds + \int_x^\infty v^2(s)ds$$

It is easy to see that X is convex. Furthermore, Tv is symmetric because $Q_2$ and v are. Next, for $v \in X$,

$$\|Tv(x)\| \le \left\| \int_x^\infty Q_2(s)ds \right\| + \int_x^\infty \|v(s)\|^2 ds$$

$$\le h\left(\int_x^\infty Q_1(s)ds\right) + \int_x^\infty v_1^2(s)ds = v_1(x)$$

if x is sufficiently large. Thus $TX \subset X$. We now proceed to show that T is continuous. To this end, let $v_n \in X$ and $v_n \to v$ in X. Then,

$$\|Tv_n(x) - Tv(x)\| \le \int_x^\infty \|v_n^2(s) - v^2(s)\| ds$$

Since $\|v_n^2 - v^2\| \le \|v_n - v\| \, \|v\| + \|v_n\| \, \|v_n - v\|$, it follows that, since $v_n$, $v \in X$, (and x is suitably large),

$$\| v_n^2(x) - v^2(x) \| \leq 2v_1(x) \, \| v_n(x) - v(x) \|$$

Hence,

$$\| Tv_n(x) - Tv(x) \| \leq 2 \, \| v_1 \|_2 \, \| \, \| v_n - v \| \, \|_2$$

which, in turn, implies that $\| Tv_n(x) - Tv(x) \| \rightarrow 0$ point-wise for every sufficiently large x. Now, since $TX \subset X$,

$$\| Tv_n(x) - Tv(x) \|^2 \leq 4v_1^2(x)$$

i.e.,

$$\int_X^\infty \| Tv_n - Tv \|^2 ds \leq 4 \int_X^\infty v_1^2 ds$$

and the Lebesgue dominated convergence theorem now gives that $\| \, \| Tv_n - Tv \| \, \|_2 \rightarrow 0$ so that T is continuous. To show that T is compact we use lemma 3.4.6. For $v \in X$,

$$\| \, \| Tv \| \, \|_2 \leq \| v_1 \|_2$$

so that (a) is satisfied. Furthermore, (b) is clear since

$$\lim_{A \rightarrow \infty} \int_{E_A} v_1^2(s) ds = 0$$

Next,

$$\| Tv(x+h) - Tv(x) \| \leq \| \int_X^{x+h} Q_2(s) ds \| + \| \int_X^{x+h} v^2(s) ds \|$$

$$\leq \| Q(x+h) - Q(x) \| + | \int_X^{x+h} v_1^2(s) ds |$$

where,

$$Q(x) \equiv \int_X^\infty Q_2(s) ds \; .$$

By hypothesis,

$$v_1(x) \geq \max\left\{ h\left(\int_x^\infty Q_1(s)ds\right), \int_x^\infty v_1^2(s)ds \right\},$$

so that, since $v_1 \in L^2$ for large x,

$$h\left(\int_x^\infty Q_1(s)ds\right) \in L^2$$

and

$$\int_x^\infty v_1^2(s)ds \in L^2.$$

But,

$$\| Q(x) \|^2 \leq \left[ h\left( \int_x^\infty Q_1(s)ds \right)\right]^2$$

so that $\|Q\| \in L^2$ for large x. Thus, by one of Lebesgue's theorems (see e.g. [Tc.1; p. 397, exercise 19]), if $\varepsilon > 0$,

$$\int_{x_o}^\infty \| Q(s+h)-Q(s) \|^2 ds < \varepsilon/4$$

whenever $|h| < \delta_1$, where $x_o$ is sufficiently large but fixed. Similarly, writing

$$V(x) \equiv \int_x^\infty v_1^2(s)ds$$

we have $V \in L^2$ and so,

$$\int_{x_1}^\infty | V(s+h) - V(s) |^2 ds < \varepsilon/4$$

for $|h| < \delta_2$, where $x_1$ is sufficiently large and fixed. Hence,

$$\int_{x_2}^\infty \| Tv(x+h) - Tv(x) \|^2 dx \leq 2\int_{x_2}^\infty \| Q(s+h) - Q(s) \|^2 ds \ +$$

$$2 \int_{x_2}^{\infty} |V(s+h) - V(s)|^2 ds < \varepsilon$$

provided $|h| < \delta = \min(\delta_1, \delta_2)$ and $x_2 = \max(x_0, x_1)$. This shows that TX is an "$L^2$-equicontinuous" family and, consequently, by lemma 3.4.6, TX is compact.

An application of Schauder's fixed point theorem now shows the existence of a fixed point for T in X which completes the proof.

**Example 1:**

We show how theorem 3.4.7 may be used to derive Taam's result [Ta.1] in the scalar case. Let $h: I \to \mathbf{R}^+$, $h \in C^1(I)$, $h(0) = 0$. Assume that $q_i \in L_1^{loc}(I)$, for i = 1, 2, and that

$$\lim_{T\to\infty} \int_x^T q_i(s)ds$$

exist and are finite, for each sufficiently large x. If

$$h\left(\int_x^{\infty} q_1(s)ds\right) \geq \left|\int_x^{\infty} q_2(s)ds\right|$$

and the equation

$$y'' + q_1(x)h'\left(\int_x^{\infty} q_1(s)ds\right)y = 0$$

is non-oscillatory on I, then

$$y'' + q_2(x)y = 0$$

is also non-oscillatory on I. This is equivalent to Taam's result for $h \in C^2(I)$ and $q_1(x) > 0$ a.e. as is not difficult to see.

One of the novelties of theorem 3.4.7 is that the existence of solutions to a non-linear matrix integral equation is reduced to the existence of solutions of a **scalar** non-linear integral equation under appropriate conditions.

**Example 2:**

On the space $S$ introduce the functional $h(A) \equiv tr(A)$, the trace of A, and let $Q_1(x) \geq 0$ (in the form sense). Now, the scalar Riccati integral equation

$$v_1(x) = tr\left(\int_x^\infty Q_1(s)ds\right) + \int_x^\infty v_1^2(s)ds$$

has a solution $v_1 \in L^2(x_0, \infty)$, some $x_0$ if the associated scalar differential equation

$$y'' + tr\, Q_1(x)y = 0,$$

is non-oscillatory on I, (as usual, it is tacitly assumed that

$$\lim_{T\to\infty} \int_x^T Q_1(s)ds$$

is a matrix with finite norm). If $Q_2 \in S$ is any matrix with

$$\left\| \int_x^\infty Q_2(s)ds \right\| \leq \int_x^\infty tr\, Q_1(s)ds$$

then the **vector** equation, $y \in \mathbf{R}^n$,

$$y'' + Q_2(x)y = 0$$

is non-oscillatory on I, (this follows from theorem 3.4.7).

**Corollary 3.4.8** **Let $Q(x) \in S$ , $x \in I$, satisfy**

$$\lim_{T\to\infty} \int_x^T Q(s)ds$$

**exists and has finite norm for sufficiently large x. If**

$$\left\| x \int_x^\infty Q(s)ds \right\| \leq 1/4$$

**for large x, then the vector equation,**

$$y'' + Q(x)y = 0$$

**is non-oscillatory on I.**

Proof: Let $Q_1(x) = \text{diag}\{1/4x^2, 0, 0, ..., 0\}$ on $[0, \infty)$. Then $\text{tr}\, Q_1(x) = 1/4x^2$ and

$$y'' + (1/4x^2)y = 0$$

is a non-oscillatory Euler equation on $[0, \infty)$. The result is now a consequence of Example 2, above.

**Remarks:** Corollary 3.4.8 is a systems analog of Wintner's classical criterion [Wi.1]. The matrix norm used here is the one defined in lemma 3.4.6. The choice of the norm is not essential in results 3.4.1 - 3.4.8 and any other (equivalent) norm may be used to derive analogous results.

**It is not known at this time whether $\|\ \ \|$ in corollary 3.4.8 may be replaced by the largest eigenvalue of the matrix in question without affecting the conclusion.**

In order to motivate a systems analog of theorem 2.5.12 we remark that, if we define the **mean-value** of the almost periodic matrix V in the natural way, viz.,

$$M\{V(x)\} = \lim_{T\to\infty} \frac{1}{T} \int_0^T V(s)ds$$

as being that constant matrix each of whose entries is the mean value of the corresponding entries of $V(x)$, we have difficulty in formulating a criterion which is, at the same time, necessary and sufficent for (3.4.1) to be oscillatory on I for every $\lambda \neq 0$.

**The natural conjecture is that (3.4.1) is oscillatory on I for every real $\lambda \neq 0$ if and only if $M\{V(x)\} = 0$.**

However, we show below that $M\{V(x)\} = 0$ is a sufficient but <u>not</u> necessary condition for the oscillation of the system (3.4.1) for every real $\lambda \neq 0$.

**Theorem 3.4.9**

**Let $V \in S$ be a non-zero almost-periodic matrix with $M\{V(x)\} = 0$. Then (3.4.1) is oscillatory on I for every real $\lambda \neq 0$.**

Proof: Let tr: $S \to \mathbf{R}$ be the positive linear functional "trace". Since V is almost-periodic, tr $V(x)$ is also a.p. and thus $\text{tr}\, M\{V(x)\} = M\{\text{tr}\, V(x)\} = 0$. Hence, the scalar equation

$$-y'' + \lambda\, \text{tr}\, V(x)\, y = 0, \qquad (3.4.7)$$

is oscillatory on **R** (and so on I) for every real $\lambda \neq 0$, by theorem 2.5.12.

Assume, if possible, that for some real $\lambda \neq 0$, (3.4.1) is non-oscillatory on I. Then, by definition, there exists $x_0 \in T$ (which we now fix) such that for every $x > x_0$, (3.4.1) is disconjugate on $[x_0, x]$. Applying lemma 3.2.8 (to a finite interval, [Ha.3]) it follows that the scalar equation (3.4.7) is disconjugate on $[x_0, x]$, for every $x > x_0$ i.e., (3.4.7) is non-oscillatory on I by Sturm theory, for such a $\lambda$, which is a contradiction.

**Remark:**

Actually, the condition on V in theorem 3.4.9 may be weakened without affecting the conclusion. Indeed, **it suffices only that, for some positive linear functional g on *S* we have that $M\{g(V(x))\} = 0$.** (For given V, the almost-periodicity of g(V(x)) follows by the characterization of such g as e.g., g(V(x)) = tr(V(x)C) where C > 0, in the form sense, and depends only on the choice of g, [Ak.1; theorem 3.1, p. 27] or by the tensor product formulation as in [Wa.1; theorem 3.1]). We leave the proof of this weaker criterion to the reader as an exercise. In this respect, see also the remark at the end of this section.

**Example 3:**

We present a two-dimensional example (n = 2) whereby $M\{V(x)\} \neq 0$ is compatible with the oscillation of (3.4.1) for every real $\lambda \neq 0$. Let

$$V(x) \equiv \begin{bmatrix} 0 & 1 \\ 1 & 0 \end{bmatrix}$$

be a constant matrix on I. The resulting system (3.4.1) is actually a two-term fourth-order o.d.e in disguise. Indeed, for $\lambda > 0$, the solution $y(x) = \text{col}\,(\sin x\sqrt{\lambda}, -\sin x\sqrt{\lambda})$ has arbitrarily large conjugate points whereas, for $\lambda < 0$, the solution $y(x) = \text{col}\,(\sin x\sqrt{-\lambda}, \sin x\sqrt{-\lambda})$ has the same property. Thus, (3.4.1) is oscillatory at $\infty$ by definition, for each real $\lambda \neq 0$.

**Note on Example 3:**

**The off-diagonal entries have non-zero mean-value here.** This motivates our principal extension below. Furthermore, *examples of every dimension n > 2 are easily constructed by* *taking appropriate direct sums of our V(x)*, if n is even, or by taking appropriate direct sums of our V(x) and a direct sum of the one-dimensional zero-matrix in the case where n is odd.

Example 3 shows that it is not sufficient for M{V(x)} to be an arbitrary non-identically-zero matrix in *S*, in order for (3.4.1) to be non-oscillatory on I for some $\lambda \neq 0$. However, if M{V(x)} is a constant multiple of the identity matrix there is such a sufficient condition.

**Theorem 3.4.10:**

**Let $V \not\equiv 0$ be an almost-periodic matrix in *S* and assume that**

$$M\{V(x)\} = m\,E$$

**where $m \in R$, and E is the identity nxn matrix. Then the necessary and sufficient condition for (3.4.1) to be oscillatory on I for every real $\lambda \neq 0$ is that $m = 0$.**

Proof: The condition is clearly sufficient on account of theorem 3.4.9.

In order to prove the necessity, we use a *tour de force*, in that, we adapt the argument used in the proof of the necessity of theorem 2.5.12 to this setting. We show that whenever $m \neq 0$, there is a value of $\lambda \neq 0$ for which (3.4.1) is non-oscillatory on I. As in the proof of said theorem we consider the single vector equation in the two parameters

$$y'' + (-\alpha E + \beta V^*(x))y = 0, \qquad x \in I,$$

where, by means of a translate by a multiple of the identity, we may assume that $M\{V^*(x)\} = 0$. Assume, as before, that $\alpha > 0$, $\beta \neq 0$. We perform the change of variables, ($y, z \in \mathbf{R}^n$),

$$y(x) = z(t) \exp(-x\sqrt{\alpha})$$

where

$$t = (1/2\sqrt{\alpha}) \exp(2x\sqrt{\alpha}).$$

Then z satisfies the equation -

$$z''(t) + \beta V^*(x) \exp(-4x\sqrt{\alpha})\, z(t) = 0.$$

We now write $f(t) \equiv \beta V^*(x) \exp(-4x\sqrt{\alpha})$ and, for simplicity, replace $V^*$ by V. Then $f \in S$.

Furthermore, for each i,j, entry of f,

$$t\int_t^\infty f_{ij}(s)ds = \frac{\beta e^{2x\sqrt{\alpha}}}{2\sqrt{\alpha}} \int_x^\infty V_{ij}(s)\exp(-2s\sqrt{\alpha})ds$$

$$= \ \ldots$$

$$= \ \beta\int_0^\infty \tau e^{-2\tau\sqrt{\alpha}}\left[\frac{1}{\tau}\int_x^{x+\tau} V_{ij}(s)ds\right]d\tau$$

by the arguments in the proof of theorem 2.5.12 which need not be represented here in detail. Now, since $M\{V_{ij}(x)\} = 0$ $(\equiv M\{V^*_{ij}(x)\})$, for each $\varepsilon > 0$ there is a $\tau_{ij}(\varepsilon) > 0$ such that

$$\left|\frac{1}{\tau}\int_x^{x+\tau} V_{ij}(s)ds\right| \le \varepsilon \tag{3.4.8}$$

uniformly for $x \in \mathbf{R}$ if $\tau \ge \tau_{ij}$. Choosing

$$\tau_0(\varepsilon) = \max_{i,j}\{\tau_{ij}(\varepsilon)\}$$

we get that (3.4.8) holds for each i, j and $\tau \ge \tau_0(\varepsilon)$. Similar changes are necessary in the rest of the proof (see theorem 2.5.12 for details). One may then show that,

$$\left|t\int_t^\infty f_{ij}(s)ds\right| \le \frac{1}{4}$$

if $\alpha > 0$ is sufficiently small, for every i, j, $1 \le i,j \le n$, i.e.,

$$\left\|t\int_t^\infty f(s)ds\right\| \le \frac{1}{4}$$

for each sufficiently large t. Hence, by corollary 3.4.8, the equation

$$z''(t) + \beta V^*(x)\exp(-4x\sqrt{\alpha})\,z(t) = 0$$

is non-oscillatory at $\infty$ and so

$$y'' + (-\alpha E + \beta V^*(x))y = 0$$

is non-oscillatory at $+\infty$ for $\alpha > 0$ sufficiently small and any $\beta \ne 0$. Since $\alpha \equiv -m\lambda$ and $\beta \equiv -\lambda$, we see that there is a $\lambda \ne 0$ such that (3.4.1) is non-oscillatory at $\infty$ for $m \ne 0$. This completes the proof of the necessity and of the theorem.

**Example 4:**

If M{V(x)} is not a constant multiple of the identity matrix E, the necessity may fail once again even if M{V(x)} is diagonal. For example, if $m_1$ and $m_2$ are constant and

$$V(x) \equiv \begin{bmatrix} m_1 & 0 \\ 0 & m_2 \end{bmatrix}$$

then, on $[0, \infty)$, (3.4.1) is oscillatory for every real $\lambda \neq 0$ provided $m_1 m_2 < 0$.

This is because, for $\lambda > 0$, the solution, (say, $m_1 > 0$), $y(x) \equiv \text{col}\,(y_1(x), y_2(x)) = \text{col}\,(0, y_2(x))$ is oscillatory since $y_2'' + \lambda|m_2|y_2 = 0$ is oscillatory while, for $\lambda < 0$ the solution $y(x) = \text{col}\,(y_1(x), 0)$ is now oscillatory since $y_1'' - \lambda m_1 y_1 = 0$ is. Examples in dimension $n > 2$ are now easily constructed.

Examples 3 and 4 show that it is perhaps difficult to get a weaker *a priori* condition on V, than the one presented in theorem 3.4.10, which still yields a necessary and sufficient condition for the system (3.4.1) to be oscillatory at $\infty$ for every real $\lambda \neq 0$.

**Remark**

As a final remark we note that if $V \in S$ and $V \not\equiv 0$ has **at least one** diagonal entry $v_{ii}(x)$ which is almost periodic with $M\{v_{ii}(x)\} = 0$, the vector differential equation

$$y'' + V(x)y = 0$$

is oscillatory at $+\infty$ (by use of an appropriate positive linear functional) and consequently,

$$-y'' + \lambda V(x)y = 0$$

is also oscillatory at $+\infty$ for every real $\lambda \neq 0$. This complements the remark following theorem 3.4.9 as V(x), here, is not necessarily a matrix consisting solely of almost-periodic entries.

## §4. Scalar Volterra-Stieltjes Integral Equations

### §4.1 Introduction

In this section we proceed to extend, as much as possible, the results in §2 in the direction of equations of the form

$$y'(x) = c + \int_{0}^{x} y(s)d\ \sigma(s), \qquad (4.1.1)_\sigma$$

for $x \in [0,\infty)$. Hence $\sigma \in BV_{loc}(0,\infty)$ (i.e., $\sigma$ is locally of bounded variation over $[0,\infty)$), and c is a constant.

The advantage in using this framework is that, basically, a corresponding theory may be derived which encompasses both ordinary differential equations and ordinary difference equations, (see e.g. [Mi.2]).

The basic theory of such equations, whose study dates back to at least W. Feller and M.G. Krein (1950's) may be found in the monograph of Atkinson [At.1] and that of Mingarelli [Mi.2].

### §4.2 Fundamental Notions.

We consolidate some of the basic results here which will be required later on.

In the sequel, $\sigma:[0,\infty) \to \mathbf{R}$ **will always be assumed to be in** $BV_{loc}(0,\infty)$ **and right-continuous.** In this case we write $\sigma \in \sum [0,\infty)$.

Furthermore **we will always assume**, for the sake of simplicity only, **that** $\sigma$ **has only a finite number of discontinuities in finite intervals.** This assumption will simplify many proofs and is, in itself, sufficient for (4.1.1) to include second-order difference equations. However, closer inspection will reveal that, in many of the results below, this assumption is unnecessary. This hypothesis will be designated by the term "etc" in the sequel.

By a **solution** of $(4.1.1)_\sigma$ is meant a function $y \in AC_{loc}[0,\infty)$ with the property that $y' \in BV_{loc}(0,\infty)$ and that y satisfies (1) on $[0, \infty)$. (it can be shown that y'(x), when generally viewed as a right-derivative, exists at each point of $[0, \infty)$) - Other notations are $y'_+(x)$ for the latter quantity.

Since y is continuous and $\sigma \in BV_{loc}(0,\infty)$, the Stieltjes integral appearing in $(4.1.1)_\sigma$ may be interpreted in the Riemann-Stieltjes sense.

It is known that initial value problems corresponding to (4.1.1) have solutions of this type which exist on $[0,\infty)$ and are unique [At.1].

We will encounter later on the, somewhat more general, equation

$$p(x)y'(x) = c + \int_0^x y(s)d\sigma(s) \tag*{$(4.2.1)_\sigma$}$$

where $p: [0,\infty) \to [0, \infty)$, and $1/p \in L_1^{loc}[0,\infty)$. In this case, a **solution** y is in $AC_{loc}([0,\infty)$ and $p(x)y'(x) \in BV_{loc}(0,\infty)$, with y satisfying $(4.2.1)_\sigma$ on $[0,\infty)$. Once again solutions of this type exist and are unique [Mi.2].

## §4.3 Disconjugate Volterra-Stieltjes integral equations

Equation (4.1.1) will be termed **disconjugate** on $[0,\infty)$ provided every non-trivial solution has at most one zero in $[0,\infty)$.

**Theorem 4.3.1. Equation (4.1.1) is disconjugate on $[0,\infty)$ if and only if there exists a solution $y(x)$ with $y(x) > 0$ for $x \in (0,\infty)$.**

Proof. Let $y_1$, $y_2$ be a pair of linearly independent solutions of (4.1.1). Then any other solution $y(x)$ is of the form $y(x) = c_1y_1(x) + c_2y_2(x)$, $c_i \in C$ are constants [At.1; Chapter 11].

Assume that (4.1.1) is disconjugate on $[0,\infty)$. If possible, assume that there is no solution y with $y(x) > 0$ on $(0,\infty)$. Then every (non-trivial) solution must have at least one zero in $(0,\infty)$. This said, let $y_3(x)$ be defined by the initial conditions $y_3(0) = 0$, $y_3'(0) \neq 0$. Then, by assumption, $y_3(x_o) = 0$ for some $x_o \in \{0,\infty)$ and so $y_3(x)$ has at least two zeros in $[0,\infty)$. This contradicts the assumption that (4.1.1) is disconjugate.

Conversely, suppose that (4.1.1) has a positive solution y in $(0,\infty)$. Let $u(x) \equiv y(x)$ and let $v(x)$ be a linearly independent solution so that $\{u(x), v(x)\}$ forms a basis for the solution space. Note that $v(x_1) = 0$, $v(x_2) = 0$ cannot both occur for $x_1 \neq x_2$ in $(0,\infty)$. For, we may assume that $u(x)\, v'(x) - u'(x)\, v(x) = 1$ [At.1, theorem 11.3.1]. Now at a zero of v, v' is, in fact, a two-sided (ordinary) derivative [At.1, theorem 11.2.2]. Hence $u(x_i)v'(x_i) = 1$, for $i = 1, 2$. But $u(x_i) > 0$ as $x_i \in (0,\infty)$. Hence $v'(x_i) > 0$ for $i = 1, 2$ while $v(x_1) = v(x_2) = 0$. Since we may assume that $x_1 < x_2$ and that v has no zeros in $(x_1, x_2)$, the latter inequalities yield a contradiction. Hence v has at most one zero in $(0,\infty)$. If $x_1 = 0$, a similar argument applies except that now $v'(x_1)$ is the right-derivative. In any event we still obtain the above contradiction. Therefore **v has at most one zero in $[0,\infty)$.**

If possible assume, on the contrary, that (4.1.1) is not disconjugate on $[0,\infty)$. Then there exists a solution $z \not\equiv 0$ and two points $x_1 < x_2$ in $[0,\infty)$ at which $z(x_1) = 0 = z(x_2)$. Fix such a solution. Then there exists constants $c_1, c_2 \in \mathbf{R}$ (as we may assume, without loss of generality, that z is real-valued) for which $z(x) = c_1u(x) + c_2v(x)$, $x \in [0,\infty)$. Clearly $c_1 \neq 0$ and $c_2 \neq 0$. Hence $u(x_i) = \alpha v(x_i)$ where $\alpha = -c_2/c_1$ and $i = 1, 2$. Now define $w(x) = \alpha v(x) - u(x)$, $x \in [0,\infty)$. Note that $w(x_i) = 0$, $i = 1, 2$. But $(\alpha v(x_i)\, v'(x_i) - u'(x_i)\, v(x_i) = 1$ and so $v(x_i)\, w'(x_i) = 1$, $i = 1, 2$. We may assume $w(0) \neq 0$. In this case, w' is, once again, a two-sided derivative at $x_i$ since $w(x_i) = 0$. Moreover $u(x_1)/u(x_2) = v(x_1)/v(x_2)$ and so $v(x_1)/v(x_2) > 0$ (as $u(x) > 0$ by hypothesis). Thus $v(x_1)$ has the same sign as $v(x_2)$. Hence $w'(x_2)$ must have the same sign. But, as before, we may assume that $x_1$ and $x_2$ are consecutive zeros of z. Since $w(x_1) = 0 = w(x_2)$ it must be the case that w also has a zero $x_3$ in $(x_1, x_2)$ (or else we are done). But $w(x_3) = 0$ means that $\alpha v(x_3) = u(x_3)$ i.e., $z(x_3) = 0$, for $x_3 \in (x_1, x_2)$, a contradiction. A similar argument applies in the event that $w(0) = 0$. Hence z can have most one zero in $[0,\infty)$ which, since z is arbitrary, implies that (4.1.1) is disconjugate on $[0,\infty)$.

We define that class $A_1^*(a, b)$, analogous to the class $A_1(a, b)$ of §2, as follows: Let $[a, b] \subset [0,\infty)$ be compact. $A_1^*(a, b) = \{\eta: [a, b] \to \mathbf{R} \mid \eta \in AC[a, b], \eta' \in BV(a, b), \eta(a) = \eta(b) = 0\}$.

Let

$$I(\eta; \sigma; a,b) = \int_a^b \{\eta'(t)^2 dt + \eta(t)^2 d\,\sigma(t)\}$$

$$= \int_a^b \eta'(t)^2 dt + \int_a^b \eta(t)^2 d\,\sigma(t)$$

for $\eta \in A_1^*(a, b)$.

**Lemma 4.3.2.** **[Mi.2, theorem 1.1.0]. If for some $\eta \not\equiv 0$ in $A_1^*(a, b)$ we have**

$$I(\eta, \sigma; a, b) \le 0,$$

**then every solution of (4.1.1) which is not a multiple of $\eta$ must have zero in (a,b).**

**Theorem 4.3.3** **Equation $(4.1.1)_\sigma$ is disconjugate on $[0,\infty)$ if and only if, for every closed and bounded subinterval [a, b] in $[0, \infty)$,**

$$I(\eta, \sigma; a, b) \ge 0$$

**for every $\eta \in A_1^*(a, b)$ with equality holding if and only if $\eta \equiv 0$.**

Proof. Let (4.1.1) be disconjugate on $[0, \infty)$. Then (by theorem 4.3.1) there exists a solution u(x) of (4.1.1) such that $u(x) > 0$ on $(0, \infty)$. So, for given [a, b], $u(x) > 0$ on (a, b). Now lemma 4.3.2 implies that for every $\eta \not\equiv 0$ in $A_1^*(a, b)$ we must have $I(\eta, \sigma; a, b) > 0$.

Conversely, let $I(\eta, \sigma; a, b)$ be positive definite on $A_1^*(a, b)$ for every $[a, b] \subset [0, \infty)$. Assume, on the contrary, that $(4.1.1)_\sigma$ is not disconjugate on $[0, \infty)$. Then there exists a solution $y \not\equiv 0$ of (4.1.1) and two points $a, b \in [0, \infty)$ such that $y(a) = y(b) = 0$. Moreover, $y \in A_1^*(a, b)$. Also, a simple calculation [Mi.2, Chapter 1], shows that

$$\int_a^b \{y'(t)^2 dt + y(t)^2 d\,\sigma(t)\} = 0,$$

i.e.

$$I(y, \sigma; a, b) = 0$$

Since I is positive definite, we must have $y \equiv 0$ on [a, b] and, by uniqueness, $y \equiv 0$ on $[0,\infty)$ which is a contradiction. Thus (4.1.1) is disconjugate on $[0,\infty)$.

**Lemma 4.3.4.** **[At.1, theorem 4.5.1]. Let y(x), $a \le x \le b$, be non-negative and continuous and $\sigma \in \Sigma\,[a, b]$ be non-decreasing. For $a \le x \le b$, let**

$$y(x) \le c_1 + c_2 \int_a^x y(s)d\,\sigma(s),$$

**where $c_1, c_2 \in R$ are constants. Then**

$$y(x) \le c_1 \exp\{c_2 \int_a^x d\sigma(s)\}$$

**for a $\le$ x $\le$ b.**

**Remark 4.1** This is a Gronwall-type inequality for Stieltjes integrals. The bound on y appearing is not generally sharp but will suffice for our purposes.For a sharp version of lemma 4.3.4 in a abstract setting see Mingarelli [Mi.1].

We now prove an extension of the Fite-Wintner-Leighton theorem alluded to in §2. In the following let p: $[0, \infty) \to [0, \infty)$ be such that $1/p \in L_1^{loc}[0, \infty)$. Let $\sigma$ be as usual.

**Theorem 4.3.5.** Let

$$\lim_{x\to\infty} \int_0^x \frac{ds}{p(s)} = +\infty \tag{4.3.1}$$

**and**

$$\lim_{x\to\infty} \sigma(x) = -\infty \tag{4.3.2}$$

**Then the equation $(4.2.1)_\sigma$ is oscillatory (see §2.2) at infinity.**

Proof. Assume, on the contrary, that $(4.2.1)_\sigma$ is non-oscillatory. Then by the Stieltjes extension of the Sturm separation theorem [Mi.2, Chapter 1] every non-trivial solution of (4.2.1) is non-oscillatory. Thus let $y \not\equiv 0$ be a non-oscillatory solution of (4.2.1). Then $y(x) \neq 0$ for all sufficiently large x and so we may take it that $y(x) > 0$ for all $x \ge x_0$, say $x_0 \in (0, \infty)$. Now let

$$v(x) = p(x)y'(x)/y(x), \qquad x \ge x_0 \tag{4.3.3}$$

Then [Mi.2, proof of theorem 2.1.1] v satisfies the integral equation

$$v(x) = v(x_0) + \sigma(x) - \sigma(x_0) - \int_{x_0}^x v^2(s)/p(s)\, ds. \tag{4.3.4}$$

for $x \ge x_0$. Thus

$$v(x) \le v(x_0) + \sigma(x) - \sigma(x_0) \qquad x \ge x_0$$

Hence (4.3.2) implies that

$$\lim_{x\to\infty} v(x) = -\infty \tag{4.3.5}$$

Thus $y'(x) \le 0$ for all $x \ge x_1 \ge x_0$. Now

$$v(x) \le -1 - \int_{x_0}^{x} v^2(x)/p(s)ds \tag{4.3.6}$$

on account of (3.4.2) if $x \ge x_2 \ge x_1$. Now (4.3.6) gives rise to

$$v(x) \le -1 - \int_{x_2}^{x} v^2(s)/p(s)ds, \qquad x \ge x_1 \tag{4.3.7}$$

Moreover, (4.3.7) may be rewritten as

$$v(x) \le -1 + \int_{x_2}^{x} |y'(s)/y(s)|\, v(s)ds$$

as $y'(x) \le 0$ for such x. Applying the Gronwall inequality to the latter we eventually obtain,

$$v(x) \le -y(x_2)/y(x), \qquad x \ge x_2.$$

i.e., (by (4.3.3)),

$$y'(x) \le -y(x_2)\cdot p(x), \qquad x \ge x_2.$$

or

$$y(x) \le y(x_2)\ [1 - \int_{x_2}^{x} \frac{ds}{p(s)}\ ], \tag{4.3.8}$$

But $y(x_2) > 0$; and (4.3.1) implies that (because of (4.3.8))

$$y(x) < 0$$

if x is sufficiently large. This is a contradiction. Therefore no non-oscillatory solution can exist and so every solution is oscillatory at infinity, i.e., (4.2.1) is oscillatory at infinity.

**Corollary 4.3.6. Let $p > 0$ a.e. on $[0,\infty)$ and satisfy (4.3.1). Let $q: [0,\infty) \to R$ satisfy $q \in L_1^{loc}[0,\infty)$ and**

$$\lim_{x \to \infty} \int_0^x q(s)ds = +\infty.$$

**Then the equation**

$$(p(x)y'(x))' + q(x)y(x) = 0 \qquad x \in [0,\infty) \qquad (4.3.9)$$

**is oscillatory at infinity.**

**Corollary 4.3.7.** **Let $c_n > 0$ for $n = -1, 0, 1, 2, \ldots$ $\{b_n\}_0^\infty$ be real sequences. If**

$$\sum_{-1}^{\infty} \frac{1}{c_n} = +\infty$$

**and**

$$\lim_{x \to \infty} \sum_{n=0}^{m} b_n = +\infty,$$

**then the second order difference equation**

$$\Delta(c_{n-1}\Delta y_{n-1}) + b_n y_n = 0$$

**is oscillatory at infinity. (Here $\Delta z_n \equiv z_{n+1} - z_n$).**

The proofs of the above corollaries follows the lines of the techniques developed in [Mi.2, Chapter 2]. Corollary 4.3.7 above is theorem 2.2.3B in [Mi.2]. Corollary 4.3.6 above is the extension of the Fite-Wintner-Leighton theorem [Sw.1] to the case when the coefficients satisfy the "**minimal conditions**", i.e., further weakening of the conditions is impossible as this would affect the actual existence and uniqueness of Carathéodory solutions of (4.3.9), see [Ev.1].

**Theorem 4.3.8.** **Let $\sigma, \nu \in \Sigma\ [0, b]$ where $b > 0$. Let $y, z$ be non-trivial solutions of $(4.1.1)_\sigma$, $(4.1.1)_\nu$ over $[0, b]$ having the same initial values (i.e., $y(0) = z(0)$, $y'(0) = z'(0)$). Then**

$$\sup_{x \in [0,b]} |y(x)-z(x)| \leq Mbc \int_0^b |(\sigma(s)-\nu(s))| \qquad (4.3.10)$$

**where $M = \sup\{|z(x)|: x \in [0,b]\}$, $c = \exp\{b \operatorname{Var}|\sigma(x)|: x \in [0,b]\}$ and the integral in (4.3.10) represents the total variation of $\sigma - \nu$ over $[0, b]$ (or $|d(\sigma - \nu)|$ is the total variation measure induced by $\sigma - \nu$).**

Proof. We have for $x \in [0,b]$,

$$y'(x) - z'(x) = \int_0^x (y(s)-z(s))d\,\sigma(s) + \int_0^x z(s)\, d\,(\sigma(s)-\nu(s)).$$

Now

$$|y(x) - z(x)| \leq \int_0^x |y'(s) - z'(s)|ds$$

$$\leq \int_0^x\int_0^t |y(\xi)-z(\xi)|\,|d\sigma(\xi)|\,dt + \int_0^x\int_0^t |z(\xi)|\,|d(\sigma(\xi))|\,dt$$

$$\leq b\int_0^x |y(\xi)-z(\xi)|\,|d\,\sigma(\xi)| + b\int_0^x |z(\xi)|\,|d(\sigma(\xi)-\nu(\xi))|.$$

(where we have interchanged the order of integration). Hence,

$$|y(x)-z(x)| \leq Mb\int_0^b |d(\sigma-\nu)\xi)| + b\int_0^x |y(\xi)-z(\xi)|\;|d\,\sigma(\xi)|$$

$$\leq Mb\int_0^b |d(\sigma-\nu)\xi)| + b\int_0^x |y(\xi)-z(\xi)|\,d\,\mu\,(\xi) \qquad (4.3.11)$$

where $\mu(x) = \int_0^x |d\,\sigma(s)|$ is non-decreasing etc... An application of lemma 4.3.4 to (4.3.11) now yields the bound

$$|y(x)-z(x)| \leq Mb\exp\{b\int_0^x d\,\mu(\xi)\}\cdot\int_0^b |d(\sigma(\xi)-\nu(\xi))|$$

from which there easily follows (4.3.10).

## § 4.4 Closure and convexity of the disconjugacy domain

We can prove many of the results of §2 in this more general setting and, at the same time, obtain their discrete counterparts.

Let $\alpha, \beta \in \mathbf{R}$ and let $D$ be the disconjugacy domain of the equation

$$y'(x) = c + \int_0^x y(s)d[-\alpha\,\sigma(s) + \beta\,\nu(s)] \tag{4.4.1}$$

where as usual $\sigma, \nu \in \Sigma\,[0,\infty)$ etc. As before, a point $(\alpha, \beta) \in D$ if and only if (4.4.1) is disconjugate on $[0,\infty)$.

**Theorem 4.4.1. The disconjugacy domain $D$ of (4.4.1) is a closed set in the usual topology of $\mathbf{R}^2$.**

Proof. Let $(\alpha_n, \beta_n) \in D$ and assume that $(\alpha_n, \beta_n) \to (\alpha, \beta)$ in $\mathbf{R}^2$. Write $\mu_n(x) \equiv -\alpha_n\,\sigma(x) + \beta_n\nu(x)$.

The proof is similar in spirit to the corresponding one in §2. Assume, on the contrary, that (4.4.1) is not disconjugate on $[0,\infty)$. Then there exists a solution $y \neq 0$ such that $y(x_1) = 0 = y(x_2)$ for some $x_1\ \ x_2 \in [0,\infty)$. Let $\delta > 0$ be some fixed number and consider (4.4.1) over $[x_1, x_2 + \delta]$.

Let $y_n(x)$ be defined by $y_n(x_1) = 0 = y(x_1)$ and $y'_n(x_1) = y'(x_1)$ and let $y_n$ satisfy

$$y_n'(x) = c + \int_0^x y_n(s)d\,\mu_n(s), \tag{4.4.2}$$

for $x \in [0, \infty)$ where $c = y'_n(0)$. Now

$$y_n'(x) = y_n'(x_1) + \int_{x_1}^x y_n(s)d\,\mu_n(s),$$

$$= y'(x_1) + \int_{x_1}^x y_n(s)\,d\,\mu_n(s) \tag{4.4.3}$$

for $x \in [x_1, x_2 + \delta]$. Applying theorem 4.3.8 over $[x_1, x_2 + \delta]$ we find

$$\sup_{x\in[x_1,x_2+\delta]} |y_n(x) - y(x)| \le O(1)\cdot \int_{x_1}^{x_2+\delta} |d(\mu_n(s) - \mu(s))|$$

However,

$$\int_{x_1}^{x_2+\delta} |d(\mu_n(s)-\mu(s))| \le \int_{x_1}^{x_2+\delta} \{|\alpha-\alpha_n|d\sigma(x)| + |\beta-\beta_n||dv(x)|\}$$

and the right-side is majorized by

$$|\alpha - \alpha_n|\cdot 0(1) + |\beta - \beta_n|\cdot 0(1)$$

where the 0(1) terms represent the total variation of $\sigma$ (and $v$) over $[x_1, x_2 + \delta]$. Hence, if $\varepsilon > 0$, we can choose n sufficiently large that

$$\sup_{x \in [x_1, x_2+\delta]} |y_n(x) - y(x)| < \varepsilon$$

for all $n \ge N$, say.

However $y(x_2) = 0$ and y changes sign around $x_2$ (since $y'(x_2) \ne 0$). Thus if $\varepsilon$ is sufficiently small (and N is appropriately chosen) $y_n(x)$ also change sign around $x_2$. Hence, for such n, $y_n$ must have a zero near $x_2$. This is impossible since $y_n$ can have at most one zero for each n and this zero is $x = x_1$, by assumption. This contradiction shows that y can have at most one zero and, since y is arbitrary, (4.4.1) must be disconjugate. Thus $(\alpha, \beta) \in D$, and this completes the proof.

An immediate consequence of theorem 4.4.1 is theorem 2.1.11. Another novelty is the **discrete** analog of said theorem.

**Corollary 4.4.2. Let $c_n > 0$, $n = -1, 0, 1, \ldots$, satisfy the condition in Corollary 4.3.7 be an infinite sequence. Consider the second-order linear difference equation with two parameters,**

$$\Delta(c_{n-1}\, \Delta y_{n-1}) + (-\alpha A_n + \beta B_n) y_n = 0 \tag{4.4.4}$$

**for $n \ge 0$. Equation (4.4.4) is said to be disconjugate at $\infty$ if there is at most one value of n such that $y_n\, y_{n+1} < 0$. Define the disconjugacy domain of (4.4.4) as usual.**

**Then $D$ is a closed set in $R^2$.**

Proof. Once again the proof relies upon choosing $\sigma$, $v$ appropriately in (4.4.1). See [Mi.2, Chapter 1 and 2] for details.

**Lemma 4.4.3** Let $\sigma, \nu \in \Sigma\ [0, \infty)$ etc. If each one of the equations $(4.1.1)_\sigma$, $(4.1.1)_\nu$ is disconjugate on $[0, \infty)$ then the equation

$$y'(x) = c + \int_0^x y(s)d[\gamma\sigma(s) + (1-\gamma)\nu(s)]$$

is also disconjugate on $[0, \infty)$ for each $\gamma \in [0, 1]$.

Proof. This is easily reconstructed once use is made of theorem 4.3.3 and the technique used in proving lemma 2.1.9. We omit the details.

**Theorem 4.4.4** Let $\sigma, \nu \in \Sigma\ [0, \infty)$ etc. Then the disconjugacy domain $D$ of (4.4.1) is a convex subset of $R^2$.

Proof. See the proof of theorem 2.1.12 except that now we use lemma 4.4.3 instead of lemma 2.1.9.

**Remark 4.2.** Note that the proof of theorem 4.3.5 does not use the assumption that $\sigma$ has a finite number of discontinuities in finite intervals and is therefore valid without this assumption. Note also that theorem 4.4.1 is also valid without the additional assumption mentioned above.

**Corollary 4.4.5.** If $D$ contains two distinct proper subspaces of $R^2$, then $D = R^2$.

**Corollary 4.4.6** Let $c_n > 0$, $n = -1, 0, 1, \ldots$ satisfy

$$\sum_0^\infty 1/c_{n-1} = \infty \,. \tag{4.4.5}$$

Let $A_n, B_n \in R$, $n = 0, 1, \ldots$ be real sequences. Then the disconjugacy domain $D$ of (4.4.4) is a convex subset of $R^2$.

**Theorem 4.4.7.** Let $\sigma \in \Sigma\ [0, \infty)$ etc. The equation

$$y'(x) = c + \int_0^x \lambda y(s)d\,\sigma(s), \qquad x \in [0, \infty) \tag{4.4.6}$$

**is disconjugate for every real value of λ if and only if $\sigma(x) \equiv$ constant ($\equiv \sigma(0)$), on $[0, \infty)$.**

Proof. This is analogous to the proof of lemma 2.1.2. If $\sigma(x) \equiv$ constant, then clearly (4.4.6) is disconjugate on $[0, \infty)$.

Conversely, assume that (4.4.6) is disconjugate on $[0, \infty)$ for every $\lambda \in \mathbf{R}$. Then by theorem 4.3.3 we must have, for every compact $[a, b] \subset [0, \infty)$, every $\eta \not\equiv 0$ in $A_1^*(a, b)$

$$I(\eta, \lambda\sigma;\ a, b) > 0$$

for every $\lambda \in \mathbf{R}$.

Now, by hypothesis, $\sigma$ is discontinuous only on a countable and nowhere dense set of points in $[0, \infty)$. Thus let $[a, b]$ be a closed and bounded interval in $[0, \infty)$ for which a, b are points of continuity of $\sigma$. Fix $\eta \not\equiv 0$ in $A_1^*(a, b)$. Then

$$\int_a^b \eta'(t)^2 dt + \lambda \int_a^b \eta(t)^2 d\,\sigma(t) > 0$$

for every $\lambda \in \mathbf{R}$. Consequently (see lemma 2.1.12),

$$\int_a^b \eta(t)^2 d\ \sigma(t) = 0 \tag{4.4.7}$$

Since $\eta$ is arbitrary it follows that (4.4.7) holds for every $\eta \in A_1^*(a, b)$. Let $\emptyset_\varepsilon(x)$ be the test functions used in the proof of lemma 2.1.2. Note that $\emptyset_\varepsilon \in A_1^*(a, b)$, for each $\varepsilon > 0$ and for every $[a, b] \subset [0, \infty)$. Hence, we must have

$$\int_a^b \emptyset_\varepsilon^2(t) d\,\sigma(t) = 0$$

i.e.,

$$0 = \int_a^{a+\varepsilon} (t-a)^2 d\sigma(t) + \int_{b-\varepsilon}^b (b-t)^2 d\sigma(t) + \int_{a+\varepsilon}^{b-\varepsilon} \varepsilon^2 d\sigma(t). \tag{4.4.8}$$

However,

$$\left|\int_a^{a+\varepsilon}(t-a)^2 d\sigma(t)\right| \le \varepsilon^2 \int_a^{a+\varepsilon} |d\sigma(t)|$$

and

$$\left|\int_{b-\varepsilon}^{b}(b-t)^2 d\sigma(t)\right| \le \varepsilon^2 \int_{b-\varepsilon}^{b} |d\sigma(t)|. \tag{4.4.9}$$

Hence

$$\frac{1}{\varepsilon^2}\left|\int_a^{a+\varepsilon}(t-a)^2 d\sigma(t)\right| = o(1) \text{ as } \varepsilon \to 0 \tag{4.4.10}$$

since $\sigma$ is continuous at $x = a$ (and a similar result holds for (4.4.9)). If $a = 0$, (4.4.10) still holds as the total variation function of a right-continuous function of bounded variation is also right-continuous. Thus, for $\varepsilon > 0$, (4.4.8) yields

$$0 = o(1) + o(1) + \sigma(b) - \sigma(a) \tag{4.4.11}$$

since $\sigma$ is continuous at each of a, b. Passing to the limit as $\varepsilon \to 0^+$ in (4.4.11) we get $\sigma(b) = \sigma(a)$.

Now in a right-neighborhood of $x = 0$, $\sigma$ must be continuous (as $\sigma$ is right-continuous at $x = 0$ and by hypothesis). The above argument then yields that

$$\sigma(x) \equiv \sigma(0), \qquad x \in [0, \delta_1)$$

where $\delta_1 > 0$ is the first discontinuity of $\sigma$. Moreover $\sigma$, is continuous in some (two-sided) neighborhood of $\delta_1$ and so the above argument once again yields that

$$\sigma(x) \equiv \sigma(0), \qquad x \in (\delta_1, \delta_2)$$

where $\delta_2$ is the next discontinuity of $\sigma$. However $\sigma$ is right-continuous at $x = \delta_1$. Hence $\sigma(\delta_1) = \sigma(0)$. We repeat this argument for $x = \delta_2, \delta_3, \ldots$ to find that for every $x \in [0, \infty)$, $\sigma(x) \equiv \sigma(0)$, and this completes the proof.

**Remark 4.3** There is no loss of generality in assuming that $\sigma(0) = 0$. For $\sigma^*(x) \equiv \sigma(x) - \sigma(0)$ shares the same properties as $\sigma$ and in fact the equations $(4.1.1)_\sigma$ and $(4.1.1)_{\sigma^*}$ are equivalent. In this case we get $\sigma^*(x) \equiv 0$ under the assumptions of the theorem.

**Corollary 4.4.8. a) We obtain lemma 2.1.2**

**b) Let $c_n > 0$ etc. and satisfy (4.4.5).**

**Let $b_n \in \mathbf{R}$ be an infinite sequence. Then**

$$\Delta(c_{n-1}\ \Delta y_{n-1}) + \lambda b_n\, y_n = 0, \quad n = 0, 1, \ldots$$

**is disconjugate at $\infty$ for every $\lambda \in \mathbf{R}$ if and only if $b_n = 0$ for each n.**

We now enunciate those results which deal with the case of second order difference equations. Most of the results in §2 are consequences of the results above.

**Theorem 4.4.9.** **Let $c_n$, $A_n$, $B_n$ be as in corollary 4.4.2. Then the disconjugacy domain of (4.4.4) is all of $\mathbf{R}^2$ if and only if $A_n = B_n = 0$ for each n.**

Proof. This is an easy consequence of corollary 4.4.8

**Corollary 4.4.10. If, for at least one value of n, we have $A_n$ or $B_n$ different from zero, then the disconjugacy domain $D$ of (4.4.4) is a proper subset of $\mathbf{R}^2$.**

**Theorem 4.4.11.** **The disconjugacy domain $D$ of (4.4.1) contains a proper subspace of the vector space $\mathbf{R}^2$ (other than the subspaces $\alpha = 0$, $\beta = 0$) if and only if there exists constants $c_1$, $c_2$, not bot zero such that the function**

$$c_1\ \sigma(x) + c_2\ v(x) \equiv \text{constant} \tag{4.4.12}$$

**for $x \in [0, \infty)$.**

Proof. Let $\sigma$, $v$ satisfy (4.4.12). Then $v(x) = c_3\sigma(x) + c_4$ for some $c_3, c_4 \in \mathbf{R}$. A glance at (4.4.1) shows that the subspace $\beta c_3 = \alpha$ is in $D$ if $c_3 \neq 0$. (If $c_3 = 0$, $v \equiv$ constant and by (4.4.12) so is $\sigma$. In this case $D = \mathbf{R}^2$ and so the result is trivial).

Conversely ,let $D$ contain the subspace $S = \{(\alpha, \beta): \beta = c\alpha, c \neq 0\}$. Then on $S$ we have that

$$y'(x) = c + \alpha \int_0^x y(s)d\,[-\sigma(x) + cv(x)]$$

must be disconjugate for every $\alpha \in \mathbf{R}$. Applying theorem 4.4.7 we get that $-\sigma(x) + cv(x) \equiv$ constant and so (4.4.12) holds.

**Remark 4.4.** Theorem 4.4.11 is the Stieltjes counterpart of theorem 2.1.7.

Note also that the **disconjugacy domain, *D* of (4.4.1) is all of $\mathbf{R}^2$ if and only if each one of $\sigma$, $\nu$ is identically constant on $[0, \infty)$**, (cf., theorem 2.1.5). Theorem 4.4.8 is the discrete analog of this result. Moreover, if $D \subset \mathbf{R}^2$ then $D$ contains at most one full ray through the origin of $\mathbf{R}^2$, (since $D$ is convex, cf., corollary 2.1.8).

**Lemma 4.4.12.** **Let $\sigma, \nu \in [0, \infty)$. If each one of the equations $(4.1.1)_\sigma$ , $(4.1.1)_\nu$ is non-oscillatory (cf., §2 for the definition) at infinity then**

$$y'(x) = c + \int_0^x y(s)d\,[\gamma\sigma(x) + (1-\gamma)\nu(s)] \qquad (4.4.13)$$

**is also non-oscillatory at infinity for each $\gamma \in [0, 1]$.**

Proof. Let $y_1(x)$ be a solution of $(4.1.1)_\nu$ such that $y_1(x) > 0$ on $[x_1, \infty)$ for some $x_1 \in [0, \infty)$. Let $y(x) = y_1(x)z(x)$ where y is a (non-trivial) solution of (4.4.13). Then $z \in AC_{loc}[x_1, \infty)$ and $z' \in BV_{loc}[x_1, \infty)$, (since y' is). Now

$$\int_{x_1}^x dy'(s) = \int_{x_1}^x y(s)d\,\mu(s)$$

where we have written $\mu(x) \equiv \gamma\sigma(x) + (1-\gamma)\nu(x)$.

$$\int_{x_1}^x y(s)d\,\mu\,(s) = \int_{x_1}^x y_1(s)z(s)d\,\mu\,(s)$$

$$= \int_{x_1}^x d(y_1'(s)z(s) \;+\; y_1(s)z'(s)$$

$$= \int_{x_1}^x \{z(s)dy_1'(s) \;+\; y_1(s)dz(s)$$

$$+\; y_1(s)dz'(s) \;+\; z'(s)dy_1(s)\}$$

$$= \int_{x_1}^x z(s)dy_1'(s) \;+\; 2\int_{x_1}^x y_1'(s)z'(s)ds \;+\; \int_{x_1}^x y_1(s)dz'(s)$$

Hence

$$\int_{x_1}^{x} y_1(s)z(s)d\,\mu(s) = \int_{x_1}^{x} z(s)y_1(s)d\,\nu(s) + 2\int_{x_1}^{x} y_1'(s)z'(s)ds + \int_{x_1}^{x} y_1(s)dz'(s),$$

i.e.,

$$\gamma\int_{x_1}^{x} y_1(s)z(s)d[\sigma(s)-\nu(s)] = 2\int_{x_1}^{x} y_1'(s)z'(s)ds + \int_{x_1}^{x} y_1(s)dz'(s).$$

Taking differentials, multiplying by $y_1(s)$ and integrating we finally obtain the equation for z, namely,

$$\int_{x_1}^{x} d(y_1^2(s)z'(s)) = \gamma\int_{x_1}^{x} y_1^2(s)z(s)d[\sigma(s)-\nu(s)].$$

i.e.,

$$y_1^2(x)z'(x) = c + \gamma\int_{x_1}^{x} y_1^2(s)z(s)d[\sigma(s) - \nu(s)] \qquad (4.4.14)_\gamma$$

where $\gamma \in [0, 1]$. Let $0 < \gamma < 1$ and write $\mu_1(x) \equiv \gamma(\sigma(s) - \nu(s))$, $\mu_2(x) \equiv \sigma(s) - \nu(s)$. We know that $(4.4.14)_1$ is non-oscillatory (since $(4.1.1)_\nu$ is and the transformation $y = y_1 z$ preserves non-oscillatory solutions). Hence, dividing $(4.4.14)_\gamma$ by $\gamma$ we find

$$\gamma^{-1}y_1^2(x)z'(x) = c' + \int_{x_1}^{x} y_1^2(s)z(s)d\,\mu_2(s). \qquad (4.4.15)$$

However, $y_1^2(x) - \gamma^{-1}y_1^2(x) = (1-\gamma^{-1})y_1^2(x) < 0$ for $x \geq x_1$. Applying the Stieltjes form of the Sturmcomparison theorem [Mi.2, Corollary 1.1.0] to the equations $(4.4.14)_1$ and (4.4.15) we obtain that (4.4.15) is non-oscillatory for each $\gamma \in (0, 1)$. Since $y(x) = y_1(x)z(x)$ is the product of non-oscillatory solutions, y itself must be non-oscillatory for each $\gamma \in (0, 1)$. The Sturm comparison theorem [Mi.2, p. 10] now implies that (4.4.13) is non-oscillatory for $\gamma \in (0, 1)$, and this completes the proof.

**Theorem 4.4.13. The non-oscillation domain (see §2), *N*, of (4.1.1) is a convex set in $R^2$.**

Proof. This follows from lemma 4.4.12 using the method of proof in theorem 2.2.2.

**Corollary 4.4.14. The non-oscillation domain of (4.4.4) is a convex subset of $R^2$ (under the assumptions in corollary 4.4.2).**

Let $\sigma, v \in \Sigma\,[0, \infty)$ and $\alpha, \beta \in \mathbf{R}$. Consider the equation

$$y'(s) = c - \int_0^x y(s)d[-\alpha\sigma(s) + \beta v(s)]\,. \tag{4.4.16}$$

**Theorem 4.4.15. Let $\sigma$, $v$ be as in (4.4.16), $\sigma$ non-decreasing for large x. If**

$$\lim_{x\to\infty} \sigma(x) = +\infty \tag{4.4.17}$$

**and**

$$-\infty < \liminf_{x\to\infty} \int_0^x v(s)\,ds \le \limsup_{x\to\infty} \int_0^x v(s)\,ds < +\infty, \tag{4.4.18}$$

**then the disconjugacy domain $D$ of (4.4.16) is contained in the right-half-plane of $R^2$ - i.e., $D \subseteq N \subseteq H^+$.**

**Remark 4.3.** Note that $\sigma$, $v$ have no assumptions regarding the distribution of their discontinuities as is usual. Moreover, theorem 4.4.15 is the Stieltjes analog of theorem 2.3.1.

**Proof of theorem 4.4.15.** Let $\alpha < 0$, $\beta \neq 0$. Write

$$v(x) = \int_0^x v(s)ds.$$

Transform (4.4.16) according to $y(x) = z(x)\exp(-\beta v(x))$. Then it can readily be shown that z satisfies the equation

$$p(x)z'(x) = c - \int_0^x z(s)d[(-\alpha)\int_0^s p(t)d\,\sigma(t) + \beta^2 \int_0^s v^2(t)p(t)dt] \tag{4.4.19}$$

where $p(x) \equiv \exp(-2\beta v(x))$. Write $\mu(x)$ for the distribution function appearing in (4.4.19). Then

$$\mu(x) \geq (-\alpha)\int_0^x p(s)d\,\sigma(s)$$

$$\geq (-\alpha)\int_{x_0}^x p(s)d\,\sigma(s) \tag{4.4.20}$$

where $x_0$ (which depends on $\beta$) is chosen so that $p(x) \geq \delta > 0$ for $x \in [x_0, \infty)$ for some $\delta$. This is certainly possible on account of (4.4.18). Fix $\beta \neq 0$. Then (4.4.20) implies that

$$\mu(x) \geq |\alpha|\ \delta\,(\sigma(x)-\sigma(x_0)).$$

Hence (by (4.4.17)),

$$\lim_{x\to\infty} \mu(x) = +\infty.$$

Furthermore,

$$\int_0^x \frac{ds}{p(s)} = \int_0^x \exp\{2\beta v(s)\}ds$$

$$\geq \delta^* x, \text{ for large } x,$$

where $\delta^*$ (which depends on $\beta$) is positive and exists on account of (4.4.18) for all sufficiently large x. Hence there holds (4.3.1). Applying theorem 4.3.5 (with $\sigma$ replaced by $-\sigma$) we find that (4.4.19) is oscillatory for each $\alpha < 0$, $\beta \neq 0$. Hence (4.4.16) is also oscillatory for such $\alpha$, $\beta$. The result follows.

**Corollary 4.4.16.** **Let $c_n$, $a_n$, $b_n$ satisfy the assumptions in corollary 4.4.2 and, in addition, $a_n \geq 0$ for all sufficiently large n. If**

$$\lim_{m\to\infty} \sum_0^m a_n = +\infty \tag{4.4.21}$$

**and**

$$-\infty < \liminf_{m\to\infty} \sum_{n=0}^m \frac{1}{c_n}\{\sum_{j=0}^n b_j\} \leq \limsup_{m\to\infty} \sum_{n=0}^m \frac{1}{c_n}\{\sum_{j=0}^n b_j\} < +\infty \tag{4.4.22}$$

**then the disconjugacy domain *D* of the equation**

$$\Delta(c_{n-1}\,\Delta y_{n-1}) + (-\alpha a_n + \beta b_n)\,y_n = 0 \tag{4.4.23}$$

**for n ≥ 0 satisfies $D \subseteq N \subseteq H^+$.**

Proof. We define step functions σ, ν as follows: Let $x_{-1} = 0 < x_1 < x_2 < ... < x_n < ...$ be the fixed partition of $[0,\infty)$ defined by

$$x_{n+1} - x_n = \frac{1}{c_n}, \quad n = -1, 0, 1, \dots .$$

Define σ, ν as step-functions whose only possible jumps are at $x = x_i$ above with $(\sigma(x_n) = \sigma(x_n + 0)$ and $\nu(x_n) = \nu(x_n+0))$,

$$\sigma(x_n) - \sigma(x_n - 0) = a_n,$$

and

$$\nu(x_n) - \nu(x_n - 0) = -b_n.$$

Then it is readily verified that the integral appearing in (4.4.18) reduces to the sums (4.4.22) and the equation (4.4.16) to (4.4.23) (see [Mi.2; Chapter 1]. We omit the details.

**Remark 4.4.** Use of the above techniques for handling differential and difference equations simultaneously makes possible discrete analogs of theorems 2.3.3-4-5 which we do not state for brevity.

**Theorem 4.4.17. Let $\sigma \in \Sigma\ [0, \infty)$ satisfy $\sigma(x_1) \neq \sigma(x_0)$ for some $x_0 < x_1$ in $[0, \infty)$. If**

$$-\infty < \liminf_{x \to \infty} v(x) \leq \limsup_{x \to \infty} v(x) < +\infty \tag{4.4.24}$$

**where**

$$v(x) \equiv \int_{x_0}^{x} \sigma(s)ds, \tag{4.4.25}$$

**and, in addition, $\sigma \notin L_2(x_0, \infty)$, then the equation**

$$y'(x) = c - \int_0^x \lambda y(s)d\,\sigma(s) \tag{4.4.26}$$

**is oscillatory at infinity for every real $\lambda \neq 0$.**

Proof. Refer to theorem 2.4.2. We use transformation $y(x) = z(x) \exp\{-\lambda v(x)\}$ which reduces (4.4.26) into an equation of the form (4.4.19) where now $\alpha = 0$. The argument is now similar to the one used in proving theorem 2.4.2 and hence we omit the details.

**Corollary 4.4.18. Let $\sigma \in \Sigma\,[0, \infty)$ have the property that v, as defined in (4.4.25), is Bohr almost-periodic. If $\sigma \notin L_2(x_0, \infty)$ then the equation (4.4.26) is oscillatory at infinity for every real $\lambda \neq 0$.**

Proof. Since v is almost-periodic, (4.4.24) holds. The conclusion follows.

**Corollary 4.4.19. Let $c_n$, $b_n$ satisfy the assumptions in corollary 4.4.2 and (4.4.22). If**

$$\lim_{m \to \infty} \sum_{n=0}^{m} \frac{1}{c_n} \left[ \sum_{j=0}^{n} b_j \right]^2 = +\infty$$

**then the equation**

$$\Delta(c_{n-1}\,\Delta y_{n-1}) + \lambda b_n y_n = 0, \qquad n = 0, 1, \dots$$

**is oscillatory at infinity for each real $\lambda \neq 0$.**

Proof. Choose $\sigma$ in theorem 4.4.17 as in the proof of corollary 4.4.16 (definition of v though). The result is now a consequence of the above considerations.

**Remark 4.5.** Let $\sigma \in \Sigma\,[0, \infty)$ not be identically constant. If $\sigma$ satisfies (4.4.24) and

$$\lim_{x \to \infty} \left\{ \frac{\sigma(x)}{x} \right\}$$

exists and is finite, then it must be equal to zero. This result is similar to theorem 2.4.3 (See §2.4 for other, similar remarks). A discrete analog of this result is easily formulated using the ideas of this section.

We may extend the results in remarks 2.9(3) to this more general setting, (See theorem 4.4.7 in this respect).

**a) Let $\sigma \in \Sigma$ [0, ∞) etc. If (4.4.6) is disconjugate for every $\lambda > 0$ ($\lambda < 0$) then $\sigma$ is non-increasing (non-decreasing) on [0, ∞).**

*From this there follows:* **If $\sigma$ strictly decreasing on some interval in [0, ∞) then there is a value of $\lambda > 0$ such that (4.4.6) is not disconjugate on [0, ∞).**

b) This implies in particular that **the bottom of the spectrum of the operator**

$$-\frac{d}{dx}\ \{y'(x) + \mu \int_0^x y(s)d\ \sigma(s)\} \tag{4.4.27}$$

**in $L_2[0,\infty)$ (or even $L_2(R)$) lies in $\{\lambda \in R: \lambda \in (-\infty, 0)\}$ if $\mu > 0$ is suitably chosen,** (see [Mi.2, Chapter 3] for the definition of (4.4.27)).

Finally, if $c_n$, $b_n$ satisfy the hypotheses in corollary 4.4.2 and **$b_n < 0$ for at least one value of n, then there is a $\lambda > 0$ such that**

$$-\Delta(c_{n-1}\ \Delta y_{n-1}) + \lambda b_n\ y_n = 0, \qquad n \geq 0$$

**is not disconjugate on [0, ∞).** There is then an associated spectral result as above.

**POSTSCRIPT:** [Added 20 May, 1888]

With regards to **Open Problem 1**, p. 16, it has been shown recently [DM.1] that if r(x) is a Besicovitch almost periodic function with M{ |r(x)| } > 0, then the necessary and sufficient condition for r(x) to be a DIPRO [see p. 16] on R is that the mean value of r(x) be equal to zero. Thus theorem 2.5.12 holds for the most general type of classical almost periodic functions -- those considered by Besicovitch [Be.1]. The results in [DM.1] have ramifications for the results of section 8.

On the basic question of Section 2.7, that is, <u>under what conditions do we have D=N for various classes of potentials</u>, we can say, using the results in [DM.1], that the analog of theorem 2.7.1 for Weyl almost-periodic functions is, in general, false. That is, there exists Weyl a.p. functions such that (2.7.2) is not disconjugate on R, but yet is oscillatory only at one end, rather than both ends of R. The example is an easy consequence of the fact that the space $L_1(R)$ is a subspace of the space of Weyl a.p. functions on R. It then follows that D and N generally split for potentials in the Weyl class (or Besicovitch class ) even though these sets are identical for potentials in the Stepanoff class ( theorem

2.7.1. ). Apparently, the failure of the Bohr uniqueness theorem ( corollary 2.8.5.) in these larger classes may account for this dichotomy.

In the paper [HKM.1], it is shown that the equation (2.1.3) admits a positive solution on R for some actually specified $\lambda$ not equal to zero, in the case when the mean value of V(x) is different from zero, as guaranteed by theorem 2.5.12.

## Bibliography


[Ad.1] N.V. Adamov, **On certain transformations not changing the integral curves of a differential equation of the first order.** (Russian) Mat. Sb. 23 (65) (1948), no. 2, 187-228. Amer. Math. Soc. Transl, Series 1, 31 (1950), 103-158.

[Ak.1] K. Akiyama, **On the maximum eigenvalue conjecture for the oscillation of second order ordinary differential systems,** M.Sc. Thesis, Department of Mathematics, University of Ottawa, 1983, x, 61 p.

[At.1] F.V. Atkinson, **Discrete and continuous boundary problems,** Academic Press, New York, (1964), xiv, 570 p.

[At.2] F.V. Atkinson, M.S.P. Eastham, J.B. McLeod, **The limit-point, limit circle nature of rapidly oscillating potentials,** University of Wisconsin, Mathematics Research Center, Technical Summary Report #1676, (1976), 30 p.

[Ba.1] J.H. Barrett, **Oscillation theory of ordinary linear differential equations,** Advances in Math., 3, (1969), 415-509.

[Be.1] A. Besicovitch, **Almost periodic functions,** Dover, New York, (1954), xiii, 180 p.

[BEM.1] G.J. Butler, L.H. Erbe, A.B. Mingarelli, **Riccati techniques and variational principles in oscillation theory for differential systems,** Trans. Amer. Math. Soc., 303, (1987), 263-282.

[Bo.1] H. Bohr, **Almost periodic functions,** Chelsea, New York, (1951), 114 p.

[Bo.2] H. Bohr, **Ein allgemeiner Satz uber die Integration eines trigonometrischen Polynoms,** Prace Mat. Fiz., 43, (1935), 1-16.

[Co.1] W. Coppel, **Disconjugacy**, Springer-Verlag, New York, Lecture Notes in Math. #220, (1971), v, 148 p.

[DM.1] A. Dzurnak and A.B. Mingarelli, **Sturm-Liouville equations with Besicovitch almost-periodicity,** Proc. Amer. Math. Soc., To appear (1989).

[Et.1] G.J. Etgen, R.T. Lewis, **Positive functionals and oscillation criteria for differential systems**, in "Optimal control and differential equations", (Proc. Conf. Univ. Oklahoma, 1977). A.B. Schwarzkopf, W.G. Kelley, S.B. Eliason Eds., Academic Press, New York, (1978), 245-275.

[Ev.1] W.N. Everitt, D. Race, **On necessary and sufficient conditions for the existence of Carathéodory solutions of ordinary differential equations,** Quaestiones Math., 2, (1978), 507-512.

[Fa.1] J. Favard, **Sur les équations différentielles linéaires à coéfficients presque périodiques**, Acta. Math., 51, (1928), 31-81.

[Fi.1] A.M. Fink, D. St.Mary, **A generalized Sturm comparison theorem and oscillation coefficients**, Monatsh. Math., 73, (1969), 207-212.

[Gl.1] I.M. Glazman, **Direct Methods of qualitative spectral analysis of singular differential operators,** Israel Program for Scientific Translation, Jerusalem, (1965)

[Ha.1] P. Hartman, **On non-oscillatory linear differential equations of the second order**, Amer. J. Math., 74, (1952), 389-400.

[Ha.2] P. Hartman, A. Wintner, **Oscillatory and non-oscillatory linear differential equations**, Amer. J. Math., 71, (1949), 627-649.

[Ha.3] P. Hartman, **Oscillation criteria for self-adjoint second order differential systems and "principal sectional curvatures"**, J. Differential Equations, 34, (1979), 326-338.

[Ha.4] P. Hartman, **Ordinary Differential Equations**, S.M. Hartman, Baltimore, (1973), xiv, 612 p.

[Ha.5] P. Hartman, **A characterization of the spectra of one-dimensional wave equations**, Amer. J. Math., 71, (1949), 915-920.

[He.1] E. Hille, **Non-oscillation theorems**, Trans. Amer. Math. Soc., 64, (1948), 234-252.

[Hl.1] J. Hale, **Ordinary Differential equations**, Wiley Interscience, New York, (1969), xvi, 332 p.

[ HKM.1] S.G. Halvorsen, M.K.Kwong and A.B. Mingarelli, **A non-oscillation theorem for second order linear equations,** in Argonne National Laboratory reprint, ANL-84-73, (1984), 119-122.

[HM.1] S.G. Halvorsen, A.B. Mingarelli, **Propriétés oscillatoires de l'équation de Sturm-Liouville à coéfficients presque périodiques**, Comptes Rendus Acad. Sci. Paris, t.299, Série 1, No. 18, (1984), 907-909.

[HM.2] S.G. Halvorsen, A.B. Mingarelli, **On the oscillation of almost-periodic Sturm-Liouville operators with an arbitrary coupling constant**, Proc. Amer. Math. Soc., 97, (1986), 269-276.

[Ho.1] K. Hoffman, R. Kunze, **Linear Algebra**, Second edition, Prentice-Hall, Englewood Cliffs N.J., (1971), viii, 407 p.

[Hw.1] S.W. Hawking, R. Penrose, **The singularities of gravitational collapse and cosmology**, Proc. Royal Soc. London Ser. A, 314, (1970), 529-548.

[Mi.1] A.B. Mingarelli, **On a Stieltjes version of Gronwall's inequality**, Proc. Amer. Math. Soc. 82, (1981), 249-251.

[Mi.2] A.B. Mingarelli, **Volterra-Stieltjes integral equations and generalized ordinary differential expressions**, Springer-Verlag, New York-Berlin, Lecture Notes in Mathematics #989, xiv, 317 p.

[Mi.3] A.B. Mingarelli, **A survey of the regular weighted Sturm-Liouville problem - The non-definite case**, in Proceedings of the Workshop on Applied Differential Equations, Tsinghua University, Beijing, The People's Republic of China, 3-7 june, 1985.; Xiao Shutie and Pu Fuquan eds., World Scientific Publishing Co. Pte. Ltd., Singapore and Philadelphia, 1986: 109-137.

[Mi.4] A.B. Mingarelli, **On the existence of conjugate points for a second order ordinary differential equation**, S.I.A.M. Journal Math. Analysis, 17, 1-6.

[Mo.1] R.A. Moore, **The least eigenvalue of Hill's equation,** J. D'Analyse Math., 5, (1956-7), 183-196.

[Mo.2] R.A. Moore, L. Markus, **Oscillation and disconjugacy for linear differential equations with almost periodic coefficients**, Acta Math., 96, (1956), 99-123.

[Mo.3] R.A. Moore, **The behavior of solutions of a linear differential equation of second order**, Pacific J. Math., 5, (1955), 125-145.

[Re.1] W.T. Reid, **Ordinary Differential equations**, John Wiley and Sons, New York, (1971).

[Re.2] W.T. Reid, **Sturmian theory of ordinary differential equations**, Springer-Verlag, New York, Applied Mathematical Sciences, #31, (1980), xv, 559 p.

[Rz.1] M. Riesz,, **Sur les ensembles compacts des fonctions sommables**, Acta Sci. Math. Szeged., 6, (1933), 136-142.

[Si.1] B. Simon, **Schrödinger Semigroups**, Bull. Amer. Math. Soc., 7, (1982), 447-526.

[Si.2] B. Simon, **Almost periodic Schrödinger operators: a review.** Adv. Applied Math. 3, (1982), 463-490.

[Sk.1] S. Stanek, **A note on the oscillation of solutions of the differential equation $y'' = \lambda q(t)y$ with a periodic coefficient**, Czech. Math. J., 29, (1979), 318-323.

[So.1] I.M. Sobol, **Investigation with the aid of polar coordinates of the asymptotic behavior of a linear differential equation of the second order**, (Russian) Mat. Sb., 28, (1951), 707-714.

[Sr.1] G.W. Stewart, **Perturbation bounds for the definite generalized eigenvalue problem,** Linear Algebra Appl., 23, (1979), 69-85.

[St.1] W. Stepanoff. **Uber einige Verallgemeinerungen der fastperiodischen Funktionen**, Math. Ann., 90, (1925) 473-492.

[Sw.1] C.A. Swanson, **Comparison and oscillation theory of linear differential equations,** Academic Press, New York, (1968).

[Ta.1] C.T. Taam, **Non-oscillatory differential equations**, Duke Math. J., 19, (1952), 493-497.

[Tc.1] E.C. Titchmarsh, **The theory of functions**, Second Edition, Oxford University Press, 1939, x, 454 p.

[Wa.1] T.J. Walters, **A characterization of positive linear functionals and oscillation criteria for matrix differential equations**, Proc. Amer. Math. Soc., 78, (1980), 198-202.

[Wi.1] A. Wintner, **On the non-existence of conjugate points**, Amer. J. Math., 73, (1951), 368-380.

[Wl.1] D. Willett, **Classification of second order linear differential equations with respect to oscillation**, Advances in Math., 1, (1967), 594-623.

[Wl.2] D. Willett, **On the oscillation of ty" + p(t)y = 0 with ∫p almost periodic**, Ann. Polon. Math., 28, (1973), 335-339.

[Ye.1] M. Yelchin, **Sur les conditions pour qu'une solution d'un système linéaire du second ordre possède deux zéros**, Comptes Rendus (Doklady) Acad. Sci. U.R.S.S., 51, (1946), 573-576.

## INDEX

## SYMBOL LIST

(The page numbers follow)

**QUICK REFERENCE GUIDE**

(To frequently quoted results)

# LECTURE NOTES IN MATHEMATICS

Edited by A. Dold and B. Eckmann

## Some general remarks on the publication of monographs and seminars

In what follows all references to monographs, are applicable also to multiauthorship volumes such as seminar notes.

§1. Lecture Notes aim to report new developments - quickly, informally, and at a high level. Monograph manuscripts should be reasonably self-contained and rounded off. Thus they may, and often will, present not only results of the author but also related work by other people. Furthermore, the manuscripts should provide sufficient motivation, examples and applications. This clearly distinguishes Lecture Notes manuscripts from journal articles which normally are very concise. Articles intended for a journal but too long to be accepted by most journals, usually do not have this "lecture notes" character. For similar reasons it is unusual for Ph.D. theses to be accepted for the Lecture Notes series.

Experience has shown that English language manuscripts achieve a much wider distribution.

§2. Manuscripts or plans for Lecture Notes volumes should be submitted either to one of the series editors or to Springer-Verlag, Heidelberg. These proposals are then refereed. A final decision concerning publication can only be made on the basis of the complete manuscripts, but a preliminary decision can usually be based on partial information: a fairly detailed outline describing the planned contents of each chapter, and an indication of the estimated length, a bibliography, and one or two sample chapters - or a first draft of the manuscript. The editors will try to make the preliminary decision as definite as they can on the basis of the available information.

§3. Lecture Notes are printed by photo-offset from typed copy delivered in camera-ready form by the authors. Springer-Verlag provides technical instructions for the preparation of manuscripts, and will also, on request, supply special staionery on which the prescribed typing area is outlined. Careful preparation of the manuscripts will help keep production time short and ensure satisfactory appearance of the finished book. Running titles are not required; if however they are considered necessary, they should be uniform in appearance. We generally advise authors not to start having their final manuscripts specially tpyed beforehand. For professionally typed manuscripts, prepared on the special stationery according to our instructions, Springer-Verlag will, if necessary, contribute towards the typing costs at a fixed rate.

The actual production of a Lecture Notes volume takes 6-8 weeks.

.../...

§4. Final manuscripts should contain at least 100 pages of mathematical text and should include

- a table of contents
- an informative introduction, perhaps with some historical remarks. It should be accessible to a reader not particularly familiar with the topic treated.
- a subject index; this is almost always genuinely helpful for the reader.

§5. Authors receive a total of 50 free copies of their volume, but no royalties. They are entitled to purchase further copies of their book for their personal use at a discount of 33.3 %, other Springer mathematics books at a discount of 20 % directly from Springer-Verlag.

Commitment to publish is made by letter of intent rather than by signing a formal contract. Springer-Verlag secures the copyright for each volume.

LECTURE NOTES

ESSENTIALS FOR THE PREPARATION
OF CAMERA-READY MANUSCRIPTS

Springer
Springer-Verlag
Berlin Heidelberg New York
London Paris Tokyo Hong Kong

The preparation of manuscripts which are to be reproduced by photo-offset require special care. Manuscripts which are submitted in technically unsuitable form will be returned to the author for retyping. There is normally no possibility of carrying out further corrections after a manuscript is given to production. Hence it is crucial that the following instructions be adhered to closely. If in doubt, please send us 1 - 2 sample pages for examination.

General. The characters must be uniformly black both within a single character and down the page. Original manuscripts are required: photocopies are acceptable only if they are sharp and without smudges.

On request, Springer-Verlag will supply special paper with the text area outlined. The standard TEXT AREA (OUTPUT SIZE if you are using a 14 point font) is 18 x 26.5 cm (7.5 x 11 inches). This will be scale-reduced to 75% in the printing process. If you are using computer typesetting, please see also the following page.

Make sure the TEXT AREA IS COMPLETELY FILLED. Set the margins so that they precisely match the outline and type right from the top to the bottom line. (Note that the page number will lie outside this area). Lines of text should not end more than three spaces inside or outside the right margin (see example on page 4).

Type on one side of the paper only.

Spacing and Headings (Monographs). Use ONE-AND-A-HALF line spacing in the text. Please leave sufficient space for the title to stand out clearly and do NOT use a new page for the beginning of subdivisons of chapters. Leave THREE LINES blank above and TWO below headings of such subdivisions.

Spacing and Headings (Proceedings). Use ONE-AND-A-HALF line spacing in the text. Do not use a new page for the beginning of subdivisons of a single paper. Leave THREE LINES blank above and TWO below headings of such subdivisions. Make sure headings of equal importance are in the same form.

The first page of each contribution should be prepared in the same way. The title should stand out clearly. We therefore recommend that the editor prepare a sample page and pass it on to the authors together with these instructions. Please take the following as an example. Begin heading 2 cm below upper edge of text area.

MATHEMATICAL STRUCTURE IN QUANTUM FIELD THEORY

John E. Robert
Mathematisches Institut, Universität Heidelberg
Im Neuenheimer Feld 288, D-6900 Heidelberg

Please leave THREE LINES blank below heading and address of the author, then continue with the actual text on the same page.

Footnotes. These should preferable be avoided. If necessary, type them in SINGLE LINE SPACING to finish exactly on the outline, and separate them from the preceding main text by a line.

**Symbols**. Anything which cannot be typed may be entered by hand in BLACK AND ONLY BLACK ink. (A fine-tipped rapidograph is suitable for this purpose; a good black ball-point will do, but a pencil will not). Do not draw straight lines by hand without a ruler (not even in fractions).

**Literature References.** These should be placed at the end of each paper or chapter, or at the end of the work, as desired. Type them with single line spacing and start each reference on a new line. Follow "Zentralblatt für Mathematik"/"Mathematical Reviews" for abbreviated titles of mathematical journals and "Bibliographic Guide for Editors and Authors (BGEA)" for chemical, biological, and physics journals. Please ensure that all references are COMPLETE and ACCURATE.

## IMPORTANT

**Pagination.** For typescript, number pages in the upper right-hand corner in LIGHT BLUE OR GREEN PENCIL ONLY. The printers will insert the final page numbers. For computer type, you may insert page numbers (1 cm above outer edge of text area).

It is safer to number pages AFTER the text has been typed and corrected. Page 1 (Arabic) should be THE FIRST PAGE OF THE ACTUAL TEXT. The Roman pagination (table of contents, preface, abstract, acknowledgements, brief introductions, etc.) will be done by Springer-Verlag.

If including running heads, these should be aligned with the inside edge of the text area while the page number is aligned with the outside edge noting that right-hand pages are odd-numbered. Running heads and page numbers appear on the same line. Normally, the running head on the left-hand page is the chapter heading and that on the right-hand page is the section heading. Running heads should not be included in proceedings contributions unless this is being done consistently by all authors.

**Corrections.** When corrections have to be made, cut the new text to fit and paste it over the old. White correction fluid may also be used.

Never make corrections or insertions in the text by hand.

If the typescript has to be marked for any reason, e.g. for provisional page numbers or to mark corrections for the typist, this can be done VERY FAINTLY with BLUE or GREEN PENCIL but NO OTHER COLOR: these colors do not appear after reproduction.

**COMPUTER-TYPESETTING.** Further, to the above instructions, please note with respect to your printout that
- the characters should be sharp and sufficiently black;
- it is not strictly necessary to use Springer's special typing paper. Any white paper of reasonable quality is acceptable.

If you are using a significantly different font size, you should modify the output size correspondingly, keeping length to breadth ratio 1 : 0.68, so that scaling down to 10 point font size, yields a text area of 13.5 x 20 cm (5 3/8 x 8 in), e.g.

Differential equations.: use output size 13.5 x 20 cm.

Differential equations.: use output size 16 x 23.5 cm.

Differential equations.: use output size 18 x 26.5 cm.

Interline spacing: 5.5 mm base-to-base for 14 point characters (standard format of 18 x 26.5 cm).
If in any doubt, please send us 1 - 2 sample pages for examination. We will be glad to give advice.

Vol. 1173: H. Delfs, M. Knebusch, Locally Semialgebraic Spaces. XVI, 329 pages. 1985.

Vol. 1174: Categories in Continuum Physics, Buffalo 1982. Seminar. Edited by F.W. Lawvere and S.H. Schanuel. V, 126 pages. 1986.

Vol. 1175: K. Mathiak, Valuations of Skew Fields and Projective Hjelmslev Spaces. VII, 116 pages. 1986.

Vol. 1176: R.R. Bruner, J.P. May, J.E. McClure, M. Steinberger, $H_\infty$ Ring Spectra and their Applications. VII, 388 pages. 1986.

Vol. 1177: Representation Theory I. Finite Dimensional Algebras. Proceedings, 1984. Edited by V. Dlab, P. Gabriel and G. Michler. XV, 340 pages. 1986.

Vol. 1178: Representation Theory II. Groups and Orders. Proceedings, 1984. Edited by V. Dlab, P. Gabriel and G. Michler. XV, 370 pages. 1986.

Vol. 1179: Shi J.-Y. The Kazhdan-Lusztig Cells in Certain Affine Weyl Groups. X, 307 pages. 1986.

Vol. 1180: R. Carmona, H. Kesten, J.B. Walsh, École d'Été de Probabilités de Saint-Flour XIV – 1984. Édité par P.L. Hennequin. X, 438 pages. 1986.

Vol. 1181: Buildings and the Geometry of Diagrams, Como 1984. Seminar. Edited by L. Rosati. VII, 277 pages. 1986.

Vol. 1182: S. Shelah, Around Classification Theory of Models. VII, 279 pages. 1986.

Vol. 1183: Algebra, Algebraic Topology and their Interactions. Proceedings, 1983. Edited by J.-E. Roos. XI, 396 pages. 1986.

Vol. 1184: W. Arendt, A. Grabosch, G. Greiner, U. Groh, H.P. Lotz, U. Moustakas, R. Nagel, F. Neubrander, U. Schlotterbeck, One-parameter Semigroups of Positive Operators. Edited by R. Nagel. X, 460 pages. 1986.

Vol. 1185: Group Theory, Beijing 1984. Proceedings. Edited by Tuan H.F. V, 403 pages. 1986.

Vol. 1186: Lyapunov Exponents. Proceedings, 1984. Edited by L. Arnold and V. Wihstutz. VI, 374 pages. 1986.

Vol. 1187: Y. Diers, Categories of Boolean Sheaves of Simple Algebras. VI, 168 pages. 1986.

Vol. 1188: Fonctions de Plusieurs Variables Complexes V. Séminaire, 1979–85. Edité par François Norguet. VI, 306 pages. 1986.

Vol. 1189: J. Lukeš, J. Malý, L. Zajíček, Fine Topology Methods in Real Analysis and Potential Theory. X, 472 pages. 1986.

Vol. 1190: Optimization and Related Fields. Proceedings, 1984. Edited by R. Conti, E. De Giorgi and F. Giannessi. VIII, 419 pages. 1986.

Vol. 1191: A.R. Its, V.Yu. Novokshenov, The Isomonodromic Deformation Method in the Theory of Painlevé Equations. IV, 313 pages. 1986.

Vol. 1192: Equadiff 6. Proceedings, 1985. Edited by J. Vosmansky and M. Zlámal. XXIII, 404 pages. 1986.

Vol. 1193: Geometrical and Statistical Aspects of Probability in Banach Spaces. Proceedings, 1985. Edited by X. Fernique, B. Heinkel, M.B. Marcus and P.A. Meyer. IV, 128 pages. 1986.

Vol. 1194: Complex Analysis and Algebraic Geometry. Proceedings, 1985. Edited by H. Grauert. VI, 235 pages. 1986.

Vol. 1195: J.M. Barbosa, A.G. Colares, Minimal Surfaces in $\mathbb{R}^3$. X, 124 pages. 1986.

Vol. 1196: E. Casas-Alvero, S. Xambó-Descamps, The Enumerative Theory of Conics after Halphen. IX, 130 pages. 1986.

Vol. 1197: Ring Theory. Proceedings, 1985. Edited by F.M.J. van Oystaeyen. V, 231 pages. 1986.

Vol. 1198: Séminaire d'Analyse, P. Lelong – P. Dolbeault – H. Skoda. Seminar 1983/84. X, 260 pages. 1986.

Vol. 1199: Analytic Theory of Continued Fractions II. Proceedings, 1985. Edited by W.J. Thron. VI, 299 pages. 1986.

Vol. 1200: V.D. Milman, G. Schechtman, Asymptotic Theory of Finite Dimensional Normed Spaces. With an Appendix by M. Gromov. VIII, 156 pages. 1986.

Vol. 1201: Curvature and Topology of Riemannian Manifolds. Proceedings, 1985. Edited by K. Shiohama, T. Sakai and T. Sunada. VII, 336 pages. 1986.

Vol. 1202: A. Dür, Möbius Functions, Incidence Algebras and Power Series Representations. XI, 134 pages. 1986.

Vol. 1203: Stochastic Processes and Their Applications. Proceedings, 1985. Edited by K. Itô and T. Hida. VI, 222 pages. 1986.

Vol. 1204: Séminaire de Probabilités XX, 1984/85. Proceedings. Edité par J. Azéma et M. Yor. V, 639 pages. 1986.

Vol. 1205: B.Z. Moroz, Analytic Arithmetic in Algebraic Number Fields. VII, 177 pages. 1986.

Vol. 1206: Probability and Analysis, Varenna (Como) 1985. Seminar. Edited by G. Letta and M. Pratelli. VIII, 280 pages. 1986.

Vol. 1207: P.H. Bérard, Spectral Geometry: Direct and Inverse Problems. With an Appendix by G. Besson. XIII, 272 pages. 1986.

Vol. 1208: S. Kaijser, J.W. Pelletier, Interpolation Functors and Duality. IV, 167 pages. 1986.

Vol. 1209: Differential Geometry, Peñíscola 1985. Proceedings. Edited by A.M. Naveira, A. Ferrández and F. Mascaró. VIII, 306 pages. 1986.

Vol. 1210: Probability Measures on Groups VIII. Proceedings, 1985. Edited by H. Heyer. X, 386 pages. 1986.

Vol. 1211: M.B. Sevryuk, Reversible Systems. V, 319 pages. 1986.

Vol. 1212: Stochastic Spatial Processes. Proceedings, 1984. Edited by P. Tautu. VIII, 311 pages. 1986.

Vol. 1213: L.G. Lewis, Jr., J.P. May, M. Steinberger, Equivariant Stable Homotopy Theory. IX, 538 pages. 1986.

Vol. 1214: Global Analysis – Studies and Applications II. Edited by Yu.G. Borisovich and Yu.E. Gliklikh. V, 275 pages. 1986.

Vol. 1215: Lectures in Probability and Statistics. Edited by G. del Pino and R. Rebolledo. V, 491 pages. 1986.

Vol. 1216: J. Kogan, Bifurcation of Extremals in Optimal Control. VIII, 106 pages. 1986.

Vol. 1217: Transformation Groups. Proceedings, 1985. Edited by S. Jackowski and K. Pawalowski. X, 396 pages. 1986.

Vol. 1218: Schrödinger Operators, Aarhus 1985. Seminar. Edited by E. Balslev. V, 222 pages. 1986.

Vol. 1219: R. Weissauer, Stabile Modulformen und Eisensteinreihen. III, 147 Seiten. 1986.

Vol. 1220: Séminaire d'Algèbre Paul Dubreil et Marie-Paule Malliavin. Proceedings, 1985. Edité par M.-P. Malliavin. IV, 200 pages. 1986.

Vol. 1221: Probability and Banach Spaces. Proceedings, 1985. Edited by J. Bastero and M. San Miguel. XI, 222 pages. 1986.

Vol. 1222: A. Katok, J.-M. Strelcyn, with the collaboration of F. Ledrappier and F. Przytycki, Invariant Manifolds, Entropy and Billiards; Smooth Maps with Singularities. VIII, 283 pages. 1986.

Vol. 1223: Differential Equations in Banach Spaces. Proceedings, 1985. Edited by A. Favini and E. Obrecht. VIII, 299 pages. 1986.

Vol. 1224: Nonlinear Diffusion Problems, Montecatini Terme 1985. Seminar. Edited by A. Fasano and M. Primicerio. VIII, 188 pages. 1986.

Vol. 1225: Inverse Problems, Montecatini Terme 1986. Seminar. Edited by G. Talenti. VIII, 204 pages. 1986.

Vol. 1226: A. Buium, Differential Function Fields and Moduli of Algebraic Varieties. IX, 146 pages. 1986.

Vol. 1227: H. Helson, The Spectral Theorem. VI, 104 pages. 1986.

Vol. 1228: Multigrid Methods II. Proceedings, 1985. Edited by W. Hackbusch and U. Trottenberg. VI, 336 pages. 1986.

Vol. 1229: O. Bratteli, Derivations, Dissipations and Group Actions on C*-algebras. IV, 277 pages. 1986.

Vol. 1230: Numerical Analysis. Proceedings, 1984. Edited by J.-P. Hennart. X, 234 pages. 1986.

Vol. 1231: E.-U. Gekeler, Drinfeld Modular Curves. XIV, 107 pages. 1986.

Vol. 1232: P.C. Schuur, Asymptotic Analysis of Soliton Problems. VIII, 180 pages. 1986.

Vol. 1233: Stability Problems for Stochastic Models. Proceedings, 1985. Edited by V.V. Kalashnikov, B. Penkov and V.M. Zolotarev. VI, 223 pages. 1986.

Vol. 1234: Combinatoire énumérative. Proceedings, 1985. Edité par G. Labelle et P. Leroux. XIV, 387 pages. 1986.

Vol. 1235: Séminaire de Théorie du Potentiel, Paris, No. 8. Directeurs: M. Brelot, G. Choquet et J. Deny. Rédacteurs: F. Hirsch et G. Mokobodzki. III, 209 pages. 1987.

Vol. 1236: Stochastic Partial Differential Equations and Applications. Proceedings, 1985. Edited by G. Da Prato and L. Tubaro. V, 257 pages. 1987.

Vol. 1237: Rational Approximation and its Applications in Mathematics and Physics. Proceedings, 1985. Edited by J. Gilewicz, M. Pindor and W. Siemaszko. XII, 350 pages. 1987.

Vol. 1238: M. Holz, K.-P. Podewski and K. Steffens, Injective Choice Functions. VI, 183 pages. 1987.

Vol. 1239: P. Vojta, Diophantine Approximations and Value Distribution Theory. X, 132 pages. 1987.

Vol. 1240: Number Theory, New York 1984–85. Seminar. Edited by D.V. Chudnovsky, G.V. Chudnovsky, H. Cohn and M.B. Nathanson. V, 324 pages. 1987.

Vol. 1241: L. Gårding, Singularities in Linear Wave Propagation. III, 125 pages. 1987.

Vol. 1242: Functional Analysis II, with Contributions by J. Hoffmann-Jørgensen et al. Edited by S. Kurepa, H. Kraljević and D. Butković. VII, 432 pages. 1987.

Vol. 1243: Non Commutative Harmonic Analysis and Lie Groups. Proceedings, 1985. Edited by J. Carmona, P. Delorme and M. Vergne. V, 309 pages. 1987.

Vol. 1244: W. Müller, Manifolds with Cusps of Rank One. XI, 158 pages. 1987.

Vol. 1245: S. Rallis, L-Functions and the Oscillator Representation. XVI, 239 pages. 1987.

Vol. 1246: Hodge Theory. Proceedings, 1985. Edited by E. Cattani, F. Guillén, A. Kaplan and F. Puerta. VII, 175 pages. 1987.

Vol. 1247: Séminaire de Probabilités XXI. Proceedings. Edité par J. Azéma, P.A. Meyer et M. Yor. IV, 579 pages. 1987.

Vol. 1248: Nonlinear Semigroups, Partial Differential Equations and Attractors. Proceedings, 1985. Edited by T.L. Gill and W.W. Zachary. IX, 185 pages. 1987.

Vol. 1249: I. van den Berg, Nonstandard Asymptotic Analysis. IX, 187 pages. 1987.

Vol. 1250: Stochastic Processes – Mathematics and Physics II. Proceedings 1985. Edited by S. Albeverio, Ph. Blanchard and L. Streit. VI, 359 pages. 1987.

Vol. 1251: Differential Geometric Methods in Mathematical Physics. Proceedings, 1985. Edited by P.L. García and A. Pérez-Rendón. VII, 300 pages. 1987.

Vol. 1252: T. Kaise, Représentations de Weil et $GL_2$ Algèbres de division et $GL_n$. VII, 203 pages. 1987.

Vol. 1253: J. Fischer, An Approach to the Selberg Trace Formula via the Selberg Zeta-Function. III, 184 pages. 1987.

Vol. 1254: S. Gelbart, I. Piatetski-Shapiro, S. Rallis. Explicit Constructions of Automorphic L-Functions. VI, 152 pages. 1987.

Vol. 1255: Differential Geometry and Differential Equations. Proceedings, 1985. Edited by C. Gu, M. Berger and R.L. Bryant. XII, 243 pages. 1987.

Vol. 1256: Pseudo-Differential Operators. Proceedings, 1986. Edited by H.O. Cordes, B. Gramsch and H. Widom. X, 479 pages. 1987.

Vol. 1257: X. Wang, On the C*-Algebras of Foliations in the Plane. V, 165 pages. 1987.

Vol. 1258: J. Weidmann, Spectral Theory of Ordinary Differential Operators. VI, 303 pages. 1987.

Vol. 1259: F. Cano Torres, Desingularization Strategies for Three-Dimensional Vector Fields. IX, 189 pages. 1987.

Vol. 1260: N.H. Pavel, Nonlinear Evolution Operators and Semigroups. VI, 285 pages. 1987.

Vol. 1261: H. Abels, Finite Presentability of S-Arithmetic Groups. Compact Presentability of Solvable Groups. VI, 178 pages. 1987.

Vol. 1262: E. Hlawka (Hrsg.), Zahlentheoretische Analysis II. Seminar, 1984–86. V, 158 Seiten. 1987.

Vol. 1263: V.L. Hansen (Ed.), Differential Geometry. Proceedings, 1985. XI, 288 pages. 1987.

Vol. 1264: Wu Wen-tsün, Rational Homotopy Type. VIII, 219 pages. 1987.

Vol. 1265: W. Van Assche, Asymptotics for Orthogonal Polynomials. VI, 201 pages. 1987.

Vol. 1266: F. Ghione, C. Peskine, E. Sernesi (Eds.), Space Curves. Proceedings, 1985. VI, 272 pages. 1987.

Vol. 1267: J. Lindenstrauss, V.D. Milman (Eds.), Geometrical Aspects of Functional Analysis. Seminar. VII, 212 pages. 1987.

Vol. 1268: S.G. Krantz (Ed.), Complex Analysis. Seminar, 1986. VII, 195 pages. 1987.

Vol. 1269: M. Shiota, Nash Manifolds. VI, 223 pages. 1987.

Vol. 1270: C. Carasso, P.-A. Raviart, D. Serre (Eds.), Nonlinear Hyperbolic Problems. Proceedings, 1986. XV, 341 pages. 1987.

Vol. 1271: A.M. Cohen, W.H. Hesselink, W.L.J. van der Kallen, J.R. Strooker (Eds.), Algebraic Groups Utrecht 1986. Proceedings. XII, 284 pages. 1987.

Vol. 1272: M.S. Livšic, L.L. Waksman, Commuting Nonselfadjoint Operators in Hilbert Space. III, 115 pages. 1987.

Vol. 1273: G.-M. Greuel, G. Trautmann (Eds.), Singularities, Representation of Algebras, and Vector Bundles. Proceedings, 1985. XIV, 383 pages. 1987.

Vol. 1274: N. C. Phillips, Equivariant K-Theory and Freeness of Group Actions on C*-Algebras. VIII, 371 pages. 1987.

Vol. 1275: C.A. Berenstein (Ed.), Complex Analysis I. Proceedings, 1985–86. XV, 331 pages. 1987.

Vol. 1276: C.A. Berenstein (Ed.), Complex Analysis II. Proceedings, 1985–86. IX, 320 pages. 1987.

Vol. 1277: C.A. Berenstein (Ed.), Complex Analysis III. Proceedings, 1985–86. X, 350 pages. 1987.

Vol. 1278: S.S. Koh (Ed.), Invariant Theory. Proceedings, 1985. V, 102 pages. 1987.

Vol. 1279: D. Ieşan, Saint-Venant's Problem. VIII, 162 Seiten. 1987.

Vol. 1280: E. Neher, Jordan Triple Systems by the Grid Approach. XII, 193 pages. 1987.

Vol. 1281: O.H. Kegel, F. Menegazzo, G. Zacher (Eds.), Group Theory. Proceedings, 1986. VII, 179 pages. 1987.

Vol. 1282: D.E. Handelman, Positive Polynomials, Convex Integral Polytopes, and a Random Walk Problem. XI, 136 pages. 1987.

Vol. 1283: S. Mardešić, J. Segal (Eds.), Geometric Topology and Shape Theory. Proceedings, 1986. V, 261 pages. 1987.

Vol. 1284: B.H. Matzat, Konstruktive Galoistheorie. X, 286 pages. 1987.

Vol. 1285: I.W. Knowles, Y. Saitō (Eds.), Differential Equations and Mathematical Physics. Proceedings, 1986. XVI, 499 pages. 1987.

Vol. 1286: H.R. Miller, D.C. Ravenel (Eds.), Algebraic Topology. Proceedings, 1986. VII, 341 pages. 1987.

Vol. 1287: E.B. Saff (Ed.), Approximation Theory, Tampa. Proceedings, 1985–1986. V, 228 pages. 1987.

Vol. 1288: Yu. L. Rodin, Generalized Analytic Functions on Riemann Surfaces. V, 128 pages, 1987.

Vol. 1289: Yu. I. Manin (Ed.), K-Theory, Arithmetic and Geometry. Seminar, 1984–1986. V, 399 pages. 1987.